STOCHASTIC SYSTEM RELIABILITY MODELING

Other books in this series

Series in modern applied mathematics—volume 5

STOCHASTIC SYSTEM RELIABILITY MODELING

Shunji Osaki

Department of Industrial and Systems Engineering
Hiroshima University, Higashi-Hiroshima 724, JAPAN

World Scientific

Published by

World Scientific Publishing Co Pte Ltd.
P.O. Box 128, Farrer Road, Singapore 9128

Library of Congress Cataloging-in-Publication Data

Osaki, Shunji.
Stochastic system reliability modeling.

(Series in modern applied mathematics; v. 5)
Includes bibliographies and index.
1. Reliability (Engineering) — Mathematical models.
2. Stochastic systems. I. Title. II. Series.
TA169.073 1985 620'.00452 85-20327
ISBN 9971-978-56-3

Reprinted 1987.

Printed in Singapore by Kyodo-Shing Loong Printing Industries Pte Ltd.

To my wife **Yoko**

and

two children **Yusuke** and **Ryosuke**

PREFACE

The purpose of this book is to present an overview of stochastic system reliability modeling for undergraduate- and graduate-students, engineers and researchers. It is assumed that the reader has some prior knowledge of basic calculus, matrix algebra, and elementary probability theory. I believe that this book could serve as a one-semester undergraduate- or graduate-level text in reliability, applied stochastic processes, stochastic operations research, and systems engineering.

The book can be divided roughly into two parts. Chapters 1-3 deals with probability theory and stochastic processes, which provide the basic ideas of applied stochastic processes, and Chapters 4-6 treats their applications to system reliability modeling. Throughout Chapters 4-6, Markov renewal processes are applied to formulating stochastic models for system reliability.

Chapter 1 develops the probability theory, which is basic and prerequisite to the following discussions. Many important concepts and theorems are presented. Such concepts and theorems are frequently used for the following discussions.

Chapter 2 presents the stochastic processes: the basic properties of stochastic processes, Poisson processes, renewal processes, Markov chains and Markov processes. Except Markov chains, the other processes are the continuous-time stochastic processes. On the other hand, Markov chains are the discrete-time stochastic processes.

Chapter 3 deals with the Markov renewal processes. A Markov renewal process combines a Markov chain and a renewal process, and is one of the most powerful tools for analyzing stochastic system reliability models since it can move one state to another and permit an arbitrary distribution of the sojourn time in a state. The fruitful results such as Markov renewal functions and stationary probabilities can be shown. Examples are presented and signal-flow graph approach for obtaining the probabilistic quantities are illustrated. Markov renewal processes with non-regeneration points are also discussed.

Chapter 4 presents stochastic models for one-unit systems. The basic measures for reliability are presented and the availability theory is developed. The well-known age and block replacement models are presented. Ordering models, which are direct extensions of replacement models, are discussed. Inspection models are also developed.

Chapter 5 develops stochastic models for two-unit redundant systems. Stochastic models for two-unit parallel and standby redundant systems are presented by applying Markov renewal processes with non-regeneration points.

Chapter 6 deals with stochastic models for fault-tolerant computing systems. The evaluation measures for performance/reliability are presented. Stochastic models for multi-processor systems are illustrated by applying Markov renewal processes and queueing theory.

Appendix A is devoted to the Laplace-Stieltjes transforms, and Appendix B is devoted to the signal-flow graphs.

I would like to thank Dr. Naoto Kaio, Hiroshima Shudo University, and Mr. Hideshi Ohshimo, Hiroshima University, who have read an earlior draft of the manuscript, corrected the mistakes, and even excellently typed the manuscript. A special thanks to Mr. Masahiko Ikemura, Hiroshima University, who has drawn all the figures. I would finally like to thank Dr. K.K. Phua, the Chief-Editor, World Scientific Publishing Co. PTE LTD, Singapore, for his kind and patient editing the book.

Shunji Osaki
Higashi-Hiroshima, Japan

CONTENTS

STOCHASTIC SYSTEM RELIABILITY MODELING

CHAPTER 1

PROBABILITY THEORY

1.1 Probability Theory

We cannot predict in advance the outcomes of tossing a coin, casting a die, tomorrow's weather, and so on. Such a trial is called a <u>random trial</u>. The theory of probability is a field of mathematics treating random phenomena. We briefly sketch the theory of probability for the later discussions.

The outcome of a random trial is call a <u>sample point</u>. The set of all possible outcomes of a trial is called a <u>sample space</u>. An <u>event</u> is a subset of a sample space. We denote a sample point by ω, an event by any capital alphabet (say, A, B, ...), and a sample space by Ω. Operations among the events can be done by adapting the operations among the sets. It is noted that a sample space Ω itself is an event and null set $\emptyset$ is an event. For any events A and B, we define:

(i) Union $A \cup B = \{\omega : \omega \in A \text{ or } \omega \in B\}$.

(ii) Intersection $A \cap B = \{\omega : \omega \in A \text{ and } \omega \in B\}$.

(iii) Complement $A^C = \{\omega : \omega \notin A\}$.
(iv) Exclusion $A \cap B = \emptyset$.

In particular, If (iv) holds, the sets A and B are said to be mutually exclusive. The sample space Ω is called the total event and $\emptyset$ the null event.

We are ready to define the probability for each event A.

(1.1.1) Definition For each event A of a sample space Ω, a number $P\{A\}$ is defined and satisfied by the following three axioms:

(i) $0 \leq P\{A\} \leq 1$.
(ii) $P\{\Omega\} = 1$.
(iii) For any sequence of events $A_1, A_2, \ldots$ that are mutually exclusive,

(1.1.2) $$P\{\bigcup_{n=1}^{\infty} A_n\} = \sum_{n=1}^{\infty} P\{A_n\}.$$

where $P\{A\}$ is called the probability of the event A.

(1.1.3) Theorem For any events A and B,

(i) $P\{A^C\} = 1 - P\{A\}$, in particular, $P\{\emptyset\} = 0$.
(ii) If $A \subset B$, then $P\{A\} \leq P\{B\}$.
(iii) $P\{A - B\} = P\{A\} - P\{A \cap B\}$, where $A - B = A \cap B^C$.
(iv) $P\{A \cup B\} = P\{A\} + P\{B\} - P\{A \cap B\}$.

For any events A and B of the sample space Ω, the

conditional probability of the event A given the event B is defined by

(1.1.4) $P\{A|B\} = P\{A \cap B\} / P\{B\}$,

if $P\{B\} > 0$. Equation (1.1.4) can be written by

(1.1.5) $P\{A \cap B\} = P\{A|B\} \cdot P\{B\}$.

If

(1.1.6) $P\{A \cap B\} = P\{A\} P\{B\}$

or

(1.1.7) $P\{A|B\} = P\{A\}$.

then the events A and B are said to be mutually independent.

1.2 Random Variables and Probability Distributions

Consider a function from a sample space Ω into the real space R. That is, consider a function f that assigns a real value to each sample point ω. In particular, we denote the image $f(A)$ that assigns a subset of R to the event A. We also denote the preimage $f^{-1}(B)$ that assigns a subset of $\omega \in \Omega$ such that $f(\omega) \in B \subset R$.

(1.2.1) Definition A random variable X on a sample space Ω is a function from Ω into the set R of real values such that the preimage of every interval of R is an event of Ω.

For instance, if we specify the interval $(a, b]$ $(a < b)$ of R, we can calculate the following probability

(1.2.2) $$P\{a < X \leq b\} = P\{\omega \in \Omega: X(\omega) \in (a, b]\}$$

from Definition (1.2.1).

The <u>probability distribution</u> or <u>distribution</u> $F(x)$ of the random variable X is defined for any real number x by

(1.2.3) $$F_X(x) = P\{X \leq x\} = P\{\omega \in \Omega: X \in (-\infty, x]\},$$

which is the probability that the random variable X is less than or equal to x.

(1.2.4) Theorem The distribution $F_X(x)$ of the random variable X is satisfied by the following properties:

(i) $F_X(x)$ is monotonely increasing and $0 \leq F_X(x) \leq 1$.
(ii) $F_X(x)$ is right continuous, i.e., $\lim_{h \downarrow 0} F_X(x+h) = F_X(x)$.
(iii) $\lim_{x \to \infty} F_X(x) = 1$ and $\lim_{x \to -\infty} F_X(x) = 0$.

A random variable X is said to be <u>discrete</u> if its set of all possible values is countable. For discrete random variables, the probability mass function $p_X(x)$ is defined by

(1.2.5) $$p_X(x) = P\{X=x\} = P\{\omega \varepsilon \Omega: X(\omega) = x\}.$$

Then the distribution of the discrete random variable X is given by

(1.2.6) $$\begin{aligned} F_X(x) &= P\{X \leq x\} \\ &= P\{\omega \varepsilon \Omega: X(\omega) \varepsilon (-\infty, x]\} \\ &= \sum_{y \leq x} p_X(y). \end{aligned}$$

A random variable X is said to be continuous if there exists a function $f_X(x)$ such that

(1.2.7) $$P\{X \varepsilon B\} = \int_B f_X(x)dx$$

where $f_X(x)$ is called the <u>probability density</u> or <u>density</u>. Then the distribution $F_X(x)$ of the continuous random variable X is given by

(1.2.8) $$F_X(x) = P\{X \leq x\} = \int_{-\infty}^{x} f_X(y)dy.$$

It is clear from the above equation that

(1.2.9) $$f_X(x) = dF_X(x)/dx.$$

For a discrete random variable the <u>expectation</u> or <u>mean</u> of the random variable X is defined by

(1.2.10) $$E[X] = \sum_x xp_X(x),$$

if the above sum exists, where x runs over all possible

values. The expectation is the weighted sum of X by the weight $p_X(x)$. For a continuous random variable X, the <u>expectation</u> or <u>mean</u> of the random variable X is defined by

(1.2.11) $$E[X] = \int_{-\infty}^{\infty} x f_X(x)dx$$

if the above integral exists.

It is quite cumbersome to distinct the random variables by the discrete one or continuous one. To overcome this, we introduce the concept of Stieltjes integrals. Combining (1.2.10) and (1.2.11), we denote the expectation of the random variable X by

(1.2.12) $$E[X] = \int_{-\infty}^{\infty} x dF_X(x)$$

which is expressed in terms of <u>Stieltjes integrals</u>. We never intend to argue Stieltjes integrals in general. In the sequel of this book, we are just interested in discrete or continuous random variables. The expectation in (1.2.12) can be interpreted by (1.2.10) if X is discrete, or by (1.2.11) if X is continuous. Throughout this book, we use Stieltjes integrals which are assumed to be the sum such as (1.2.10) if X is discrete, or the integral such as (1.2.11) if X is continuous.

If X is a random variable, X^2 is also a random variable. The expectation of X^2 is given by

(1.2.13) $$E[X^2] = \int_{-\infty}^{\infty} x^2 dF_X(x)$$

which is called the <u>second moment</u> of X about the origin. The <u>variance</u> of the random variable X is given by

$$(1.2.14)\qquad \mathrm{Var}(X) = E[(X - E[X])^2]$$
$$= \int_{-\infty}^{\infty} (x - E[X])^2 dF_X(x)$$
$$= \int_{-\infty}^{\infty} x^2 dF_X(x) - E[X]^2$$

which is the second moment of the random variable X about its mean $E[X]$. The <u>standard deviation</u> of the random variable X is defined by the positive root of the variance, $\sqrt{\mathrm{Var}(X)}$. The standardized random variable Y corresponding to X is defined by

$$(1.2.15)\qquad Y = (X - E[X]) / \sqrt{\mathrm{Var}(X)},$$

where $E[Y] = 0$ and $\mathrm{Var}(Y) = 1$.

In general, the <u>n^{th} moment</u> of the random variables X about the origin is defined by

$$(1.2.16)\qquad E[X^n] = \int_{-\infty}^{\infty} x^n dF_X(x) \qquad (n = 1, 2, \ldots),$$

if it exists. Similarly, the nth moment of the random variable X about its mean is defined by

$$(1.2.17)\qquad E[(X - E[X])^n] = \int_{-\infty}^{\infty} (x - E[X])^n dF_X(x) \qquad (n = 1, 2, \ldots),$$

if it exists. It is clear that

$$(1.2.18)\qquad E[(X - E[X])^2] = \mathrm{Var}(X) = E[X^2] - E[X]^2,$$

$$(1.2.19)\qquad E[(X - E[X])^3] = E[X^3] - 3E[X^2]E[X] + 2E[X]^3,$$

and

$$(1.2.20)\quad E[(X - E[X])^4] = E[X^4] - 4E[X^3]E[X] + 6E[X^2]E[X]^2 - 3E[X]^4.$$

It is sometimes convenient to define the dimensionless moments about its mean,

$$(1.2.21)\quad a_n = E[(X - E[X])^n] / (\sqrt{Var(X)})^n \qquad (n = 1, 2, \ldots$$

Noting that $E[X - E[X]] = 0$ and equation (1.2.18), we have that $a_1 = 0$ and $a_2 = 1$. In addition, a_3 is called the moment coeffecient of skewness, where skewness is the degree of asymmetry, and a_4 is called the moment coefficient of kurtosis, where kurtosis is the degree of peakedness.

The integral $\int_a^b g(x)dF(x)$ is called a Lebesgue-Stieltjes integral which is frequently used for the later discussions. We give a useful formula for integrating by parts for the Lebesgue-Stieltjes integral. If $g(x)$ is a bounded function with continuous derivative $g'(x)$, then

$$(1.2.22)\ \int_a^b g(x)dF(x) = g(b)F(b) - g(a)F(a) - \int_a^b g'(x)F(x)dx.$$

A simple application of the above formula is the following: We consider a non-negative random variable X with $F(0) = 0$. If we assume that $g(x) = x^n$, $F(x) = F_X(x)$, and $a = 0$, we have

$$(1.2.23)\ \int_0^b x^n dF_X(x)$$

$$= b^n F_X(b) - n\int_0^b x^{n-1} F_X(x)dx$$

$$= -b^n[1 - F_X(b)] + n\int_0^b x^{n-1}[1 - F_X(x)]dx.$$

Noting that $-b^n[1 - F_X(b)] \to 0$ as $b \to \infty$, we have

(1.2.24) $$\int_0^\infty x^n dF_X(x) = n \int_0^\infty x^{n-1}[1 - F_X(x)]dx.$$

In particular, the expectation is

(1.2.25) $$\int_0^\infty x dF_X(x) = \int_0^\infty [1 - F_X(x)]dx$$

for $n = 1$, and the second moment about the origin is

(1.2.26) $$\int_0^\infty x^2 dF_X(x) = 2\int_0^\infty x[1 - F_X(x)]dx$$

for $n = 2$. The formula (1.2.24) will be used for calculating the nth moment about the origin for non-negative random variables with $F(0) = 0$.

The characteristic function of the random variable X is defined by

(1.2.27) $$E[e^{iux}] = \int_{-\infty}^\infty e^{iux} dF_X(x)$$

if it exists, where $i = \sqrt{-1}$ is an imaginary unit.

(1.2.28) Theorem The probability distribution is one-to-one correspondent with its characteristic function.

Theorem (1.2.28) insists that the probability distribution specifies its characteristic function uniquely, and vice versa.

In elementary statistics, instead of the characteristic function, the moment generating function of the random variable X is defined by

(1.2.29) $$M_X(\theta) = \int_{-\infty}^{\infty} e^{\theta x} dF_X(x),$$

if it exists. The above integral is just a real integral. However, the integral (1.2.27) is a complex integral.

In reliability theory, we frequently use the Laplace-Stieltjes transform of the non-negative random variable X defined by

(1.2.30) $$F_X^*(s) = \int_0^{\infty} e^{-sx} dF_X(x),$$

if it exists. In general, the Laplace-Stieltjes transforms can be adopted for the non-negative random variables. Since the reliability theory is concerned with the non-negative random variables that represent the real time $t \geq 0$, the Laplace-Stieltjes transforms can be used in reliability theory.

Theorem (1.2.28) can be rewritten by changing the characteristic function for the moment generating function or Laplace-Stieltjes transform. In the sequel of this book, we are mainly interested in reliability models whose domains are, of course, time axes. Once the Laplace-Stieltjes transform can be derived, we can invert its Laplace-Stieltjes transform. The inverse transform is a function of time. However, the analytical inversion will be very difficult or impossible except the simplest cases. We have to adopt the numerical inversion of the Laplace-Stieltjes transform for most cases. Of course, the asymptotic behavior of the time $t \to \infty$ can be easily derived from the Laplace-Stieltjes transforms and Tauberian theorems. Appendix A is devoted to the Laplace-Stieltjes transforms and their basic properties.

The nth moment of the random variable X about the origin can be easily derived from its characteristic function as follows:

$$(1.2.31) \qquad E[X^n] = \frac{1}{i^n} \left. \frac{d^n \phi_X(u)}{du^n} \right|_{u=0} .$$

The similar formula of (1.2.31) can be easily derived for the moment generating function or Laplace-Stieltjes transform.

Assume that we perform a trial whose outcome can be classified as either a "success" or as a "failure". Such a trial is called a Bernoulli trial. The following three discrete distributions are derived from the Bernoulli trials.

(1.2.32) Example The probability of k successes out of n trials is given by

$$(1.2.33) \qquad p_X(k) = \binom{n}{k} p^k q^{n-k} \qquad (k = 0, 1, 2, \ldots, n),$$

which is the probability mass function of the binomial distribution, where p and $q = 1 - p$ are the probabilities of success and failure, respectively. The mean and variance are given by $E[X] = np$ and $Var(X) = npq$, respectively.

(1.2.34) Example The probability of the necessary trials of failure to the first success is given by

$$(1.2.35) \qquad p_X(x) = pq^{x-1} \qquad (x = 1, 2, \ldots),$$

which is the probability mass function of the geometric distribution. The mean, variance, and characteristic function are given by $E[X] = 1/p$, $Var(X) = q/p^2$, and $\phi_X(u) = pe^{iu}/[1 - qe^{iu}]$, respectively.

(1.2.36) Example The third distribution is concerned with the probability of the necessary trials of failure to the first r successes $(r \geq 1)$:

$$(1.2.37) \qquad p_X(x) = \binom{x-1}{x-r} p^r q^{x-r}$$

$$= \binom{-r}{x-r} p^r (-q)^{x-r} \qquad (x = r, r+1, \ldots),$$

which is the probability mass function of the negative binomial distribution or Pascal distribution. The mean, variance, and characteristic function are given by $E[X] = r/p$, $Var(X) = rq/p^2$, and $\phi_X(u) = \{pe^{iu}/[1 - qe^{iu}]\}^r$, respectively. Note that the sum of the independent and identically distributed r geometric random variables obeys the negative binomial distribution. This fact can be easily understood by the relationship between their characteristic functions.

Table 1.2.1 shows the typical discrete distributions, probability mass functions, means, variances, and characteristic functions. In particular, the Poisson distribution will be frequently used as far as the Poisson process is concerned. Table 1.2.2 also shows the typical continuous distributions, densities, means, variances, and characteristic functions. The typical continuous and discrete distributions in reliability theory will be explained in detail in Section 1.4.

Table 1.2.1 Discrete Distributions

Discrete Distribution	Domain Parameters	Probability Mass Functions	Mean	Variance	Characteristic Functions
Binomial	$x = 0, 1, 2, \ldots, n$ $0 < p < 1,\ q = 1 - p$ n : Positive Integer	$\binom{n}{x} p^x q^{n-x}$	np	npq	$(pe^{iu} + q)^n$
Poisson	$x = 0, 1, 2, \ldots$ $\lambda > 0$	$e^{-\lambda} \frac{\lambda^x}{x!}$	λ	λ	$\exp[\lambda(e^{iu} - 1)]$
Geometric	$x = 1, 2, 3, \ldots$ $0 < p < 1,\ q = 1 - p$	pq^{x-1}	$\frac{1}{p}$	$\frac{q}{p^2}$	$\frac{pe^{iu}}{1 - qe^{iu}}$
Negative Binomial (Pascal)	$x = r, r+1, r+2, \ldots$ $0 < p < 1,\ q = 1 - p$ r : Positive Integer	$\binom{x-1}{x-r} p^r q^{x-r}$ $= \binom{-r}{x-r} p^r (-q)^{x-r}$	$\frac{r}{p}$	$\frac{rq}{p^2}$	$\left(\frac{pe^{iu}}{1 - qe^{iu}}\right)^r$
Uniform	$x = 0, 1, 2, \ldots, n$ n : Positive Integer	$\frac{1}{n + 1}$	$\frac{n}{2}$	$\frac{n^2 + 2n}{12}$	$\frac{1}{n + 1} \cdot \frac{1 - e^{iu(n+1)}}{1 - e^{iu}}$

Table 1.2.2 Continuous Distributions

Continuous Distribution	Domain Parameters	Density	Mean	Variance	Characteristic Function
Exponential	$t \geq 0$ $\lambda > 0$	$\lambda e^{-\lambda t}$	$1/\lambda$	$1/\lambda^2$	$\frac{\lambda}{\lambda - iu}$
Gamma	$t \geq 0$ $\lambda > 0,\ n > 0$	$\frac{\lambda e^{-\lambda t}(\lambda t)^{n-1}}{\Gamma(n)}$	n/λ	n/λ^2	$(\frac{\lambda}{\lambda - iu})^n$
Uniform	$a \leq t \leq b$ $a < b$	$\frac{1}{b - a}$	$\frac{a + b}{2}$	$\frac{(b - a)^2}{12}$	$\frac{e^{iub} - e^{iua}}{iu(b - a)}$
Normal	$-\infty < t < \infty$ $\mu,\ \sigma > 0$	$\frac{1}{\sqrt{2\pi}\sigma} e^{-\frac{(t-\mu)^2}{2\sigma^2}}$	μ	σ^2	$\exp(iu\mu - \frac{u^2\sigma^2}{2})$
Log Normal	$t \geq 0$ $-\infty < \mu < \infty$ $\sigma > 0$	$\frac{1}{t\sqrt{2\pi}\sigma} e^{-\frac{(\log t-\mu)^2}{2\sigma^2}}$	$e^{(\mu+\frac{\sigma^2}{2})}$	$e^{2\mu+\sigma^2}(e^{\sigma^2} - 1)$	

1.3 Joint Distributions and Limit Theorems

The random variables introduced in the preceding section was a function that assigns a real value (i.e., a one-dimensional space) to each event. If a two-dimensional real space can be considered, the corresponding bivariate distribution is defined by

$$F_{XY}(x,y) \doteq P\{X \leq x,\ Y \leq y\}. \tag{1.3.1}$$

Such a distribution is called the joint distribution of two random variables X and Y. The marginal distribution of the random variable X is defined by

$$\begin{aligned} F_X(x) &= P\{X \leq x\} \\ &= \lim_{y\to\infty} P\{X \leq x,\ Y \leq y\} \\ &= F_{XY}(x,\infty). \end{aligned} \tag{1.3.2}$$

The marginal distribution of Y is similarly defined by

$$F_Y(y) = F_{XY}(\infty,y). \tag{1.3.3}$$

The random variables X and Y are said to be independent if

$$F_{XY}(x,y) = F_X(x)F_Y(y) \tag{1.3.4}$$

for all x and y.

If the two random variables are discrete, the probability mass function of the two random variables is defined by

(1.3.5) $$p_{XY}(x,y) = P\{X = x, Y = y\},$$

and the corresponding distribution is given by

(1.3.6) $$F_{XY}(x,y) = \sum_{a \le x} \sum_{b \le y} p_{XY}(a,b).$$

In addition, if

(1.3.7) $$p_{XY}(x,y) = p_X(x)p_Y(y)$$

for all x and y, then the two random variables X and Y are <u>independent</u>, where

(1.3.8) $$p_X(x) = \sum_y p_{XY}(x,y)$$

is the marginal probability mass function of X, and $P_Y(y)$ is the marginal probability mass function of Y, which is similarly defined.

If the two random variables are continuous, the distribution is given by

(1.3.9) $$F_{XY}(x,y) = \int_{-\infty}^{x} \int_{-\infty}^{y} f_{XY}(x,y)dxdy$$

where $f_{XY}(x,y)$ is called the <u>joint density</u> of X and Y. In addition, if

(1.3.10) $$f_{XY}(x,y) = f_X(x)f_Y(y)$$

for all x and y, then the two random variables X and Y are <u>independent</u>, where

(1.3.11) $$f_X(x) = \int_{-\infty}^{\infty} f_{XY}(x,y)dy$$

is the marginal density of X, and $f_Y(y)$ is the marginal density of Y, which is similarly defined.

(1.3.12) Example The bivariate exponential distribution of two non-negative random variables X and Y is given by its survival probability

(1.3.13) $$\begin{aligned}\bar{F}_{XY}(x,y) &= P\{X > x, Y > y\} \\ &= \exp[-\lambda_1 x - \lambda_2 y - \lambda_{12}\max(x,y)],\end{aligned}$$

where $x \geq 0$, $y \geq 0$, $\lambda_1 \geq 0$, $\lambda_2 \geq 0$, and $\lambda_{12} \geq 0$. The marginal distributions of X and Y are

(1.3.14) $$F_X(x) = 1 - \exp[-(\lambda_1 + \lambda_{12})x],$$

(1.3.15) $$F_Y(y) = 1 - \exp[-(\lambda_2 + \lambda_{12})y],$$

which are exponential distributions. If $\lambda_{12} = 0$, then

(1.3.16) $$F_{XY}(x,y) = F_X(x)F_Y(y)$$

for all x and y, which implies that X and Y are independent. On the other hand, if $\lambda_{12} > 0$, then X and Y are dependent.

Consider the expectation or mean of $X + Y$. If the two random variables are discrete, then

$$(1.3.17) \qquad E[X+Y] = \sum_x \sum_y (x+y)p_{XY}(x,y)$$
$$= \sum_x xp_X(x) + \sum_y yp_Y(y)$$
$$= E[X] + E[Y].$$

The above identity can be similarly verified if the two random variables are continuous. In general, if k_1 and k_2 are constants, then

$$(1.3.18) \qquad E[k_1x + k_2y] = k_1E[X] + k_2E[Y].$$

The variance of $X + Y$ is given by

$$(1.3.19) \qquad Var(X+Y) = E[((X+Y) - E[X+Y])^2]$$
$$= Var(X) + Var(Y) + 2(E[XY] - E[X]E[Y])$$
$$= Var(X) + Var(Y) + 2Cov(X,Y),$$

where

$$(1.3.20) \qquad Cov(X,Y) = E[(X - E[X])(Y - E[Y])]$$
$$= E[XY] - E[X]E[Y]$$

is called the <u>covariance</u> of the two random variables X and Y. The <u>correlation coefficient</u> of X and Y is defined by

$$(1.3.21) \qquad \rho(X,Y) = Cov(X,Y) / \sqrt{Var(X)Var(Y)},$$

which is the dimensionless moment such that $-1 \leq \rho(X,Y) \leq 1$.

It is clear that, if X and Y are independent, then $Cov(X,Y) = 0$ and $\rho(X,Y) = 0$. However, it is not true that the inverse holds (i.e., $Cov(X,Y) = 0$ implies the independence of X and Y).

If X and Y are independent with respective distributions $F_X(x)$ and $F_Y(y)$, the distribution of $X + Y$ is given by

$$(1.3.22) \qquad F_{X+Y}(a) = P\{X+Y \leq a\} = \int_{-\infty}^{\infty} F_X(a - y)dF_Y(y) = \int_{-\infty}^{\infty} F_Y(a - x)dF_X(x)$$

which is called the <u>convolution</u> of $F_X(x)$ and $F_Y(y)$. In particular, we denote $F_X * F_Y(a) = F_{X+Y}(a)$ for the convolution. Let the characteristic functions of X, Y, and $X + Y$ by $\phi_X(u)$, $\phi_Y(u)$, and $\phi_{X+Y}(u)$, respectively. Then

$$(1.3.23) \qquad \phi_{X+Y}(u) = \int_{-\infty}^{\infty} e^{iua} \frac{d}{da}[\int_{-\infty}^{\infty} F_X(a-y)dF_Y(y)]da = \int_{-\infty}^{\infty} e^{iu(a-y)} \frac{d}{da}F_X(a-y)da \int_{-\infty}^{\infty} e^{iuy}dF_Y(y) = \phi_X(u)\phi_Y(u) .$$

That is, the characteristic function of $X + Y$ is the product of the characteristic functions of X and Y. Of course, if we can change the characteristic function for the moment generating function or Laplace-Stieltjes transform, the same property holds.

We can easily extend the n-dimensional distribution:

(1.3.24) $$F_{X_1 \dots X_n}(x_1, x_2, \dots, x_n) = P\{X_1 \le x_1, X_2 \le x_2, \dots, X_n \le x_n\} .$$

The marginal distribution of X_i is defined by

(1.3.25) $$F_{X_i}(x_i) = \lim_{\substack{x_j \to \infty \\ j \ne i}} F_{X_1 \dots X_n}(x_1, x_2, \dots, x_n) .$$

The n random variables are said to be <u>independent</u> if

(1.3.26) $$F_{X_1 \dots X_n}(x_1, x_2, \dots, x_n) = F_{X_1}(x_1) F_{X_2}(x_2) \dots F_{X_n}(x_n)$$

for all x_1, x_2, x_3, ..., x_n.

The expectation or mean of $k_1X_1 + k_2X_2 + \dots + k_nX_n$ is

(1.3.27) $$E[k_1X_1 + k_2X_2 + \dots + k_nX_n]$$
$$= k_1E[X_1] + k_2E[X_2] + \dots + k_nE[X_n] ,$$

where k_1, k_2, ..., k_n are constants. The variance of $X_1 + X_2 + \dots + X_n$ is

(1.3.28) $$\mathrm{Var}(X_1 + X_2 + \dots + X_n)$$
$$= \sum_i \sum_j \mathrm{Cov}(X_i, X_j)$$
$$= \sum_{i=1}^{n} \mathrm{Var}(X_i) + 2 \sum_{i>j} \sum \mathrm{Cov}(X_i, X_j) ,$$

where it is noted that $\mathrm{Cov}(X_i, X_i) = \mathrm{Var}(X_i)$.

In the sequel, we are interested in the convolution of

the independent and identically distributed random variables with distribution $F(x)$, say. Then we denote

(1.3.29) $$F^{(2)}(x) = F*F(x)$$

and, in general,

(1.3.30) $$F^{(n)}(x) = F*F^{(n-1)}(x) \qquad (n = 1, 2, \ldots),$$

which is called the n-fold convolution of $F(x)$ with itself, where $F^{(0)}(x) = 1(t)$ is a unit function at $x = 0$. The notation $F^{(n)}(x)$ will be frequently used in renewal theory and Markov renewal theory.

The following limit theorems are of great use in the later discussions.

(1.3.31) Theorem (Strong Law of Large Numbers) If X_1, X_2, ..., X_n are independent and identically distributed random variables with finite mean μ and variance σ^2, then

(1.3.32) $$P\{\lim_{n\to\infty}(X_1 + X_2 + \ldots + X_n)/n = \mu\} = 1.$$

This theorem insists that the sample mean $(X_1+X_2+\ldots+X_n)/n$ tends to the population mean μ with probability 1 as $n \to \infty$.

(1.3.33) Theorem (Central Limit Theorem) If X_1, X_2, ..., X_n are independent and identically distributed random variables with finite mean μ and variance σ^2, then

(1.3.34) $$P\{(X_1+X_2+\ldots+X_n - n\mu)/\sigma\sqrt{n} \le k\}$$
$$\to (1/\sqrt{2\pi})\int_{-\infty}^{k} \exp(-x^2/2)dx$$

as $n \to \infty$.

Let $Y = (X_1 + X_2 + \ldots + X_n)/n$ denote the sample mean. Then

(1.3.35) $$E[Y] = E[(X_1+X_2+\ldots+X_n)/n] = \mu ,$$

(1.3.36) $$\mathrm{Var}(Y) = \mathrm{Var}((X_1+X_2+\ldots+X_n)/n) = \sigma^2/n .$$

The standardized random variables of Y is given by

(1.3.37) $$(Y - E[Y])/\sqrt{\mathrm{Var}(Y)} = (X_1+X_2+\ldots+X_n - n\mu)/\sigma\sqrt{n} .$$

From this fact, the central limit theorem insists that the standardized random variable of the sample mean Y tends to the normal distribution $N(0,1)$ with probability 1 as $n \to \infty$. Note that the notation $N(0,1)$ means the normal distribution with mean 0 and variance 1.

(1.3.38) Example If $X_1, X_2, \ldots, X_r$ are independent and identically distributed geometric random variables with parameter p (see Table 1.2.1), the characteristic function of the distribution of the random variable $S_r = X_1 + X_2 + \ldots + X_r$ is

(1.3.39) $$\phi_{S_r}(u) = \{p/[1 - qe^{iu}]\}^r ,$$

which is the characteristic function of the negative binomial distribution. That is, the r-hold convolution $F^{(r)}(x)$ of $F(x)$ with itself is the negative binomial distribution. This fact has been described in Section 1.2.

(1.3.40) Example If $X_1, X_2, \ldots, X_n$ are independent and identically distributed exponential random variables with parameter λ and $S_n = X_1 + X_2 + \ldots + X_n$, then the Laplace-Stieltjes transform of $P\{S_n \leq t\}$ is

$$(1.3.41) \qquad F^*_{S_n}(s) = [\lambda/(s + \lambda)]^n ,$$

which is the Laplace-Stieltjes transform of the gamma distribution (see Appendix A). That is, the n-fold convolution $F^{(n)}(t)$ of $F(t)$ with itself is the gamma distribution (see Table 1.2.2).

1.4 Lifetime Distributions in Reliability Theory

1.4.1 Continuous Distributions

We have introduced some distributions in the preceding two sections. In this section, we focus on the lifetime distributions arising in reliability theory. First, we restrict ourselves to the continuous distributions in Subsection 1.4.1. Secondly, we discuss the discrete distributions in Subsection 1.4.2.

The distribution of the non-negative random variable X is given by

$$(1.4.1) \qquad F(t) = P\{X \leq t\} \quad (t \geq 0),$$

which describes the lifetime distribution of an item such as the material, part, component, or system. The survival probability of X is given by

(1.4.2) $$\bar{F}(t) = 1 - F(t) = P\{X > t\} \quad (t \geq 0),$$

which is the probability that the item survives up to time t and is called the <u>reliability function</u> of the item. The probability density of the random variable X is assumed to exist and is given by

(1.4.3) $$f(t) = dF(t)/dt \quad (t \geq 0),$$

since the continuous random variable is concerned.

The <u>failure rate</u> or <u>hazard rate</u> is defined by

(1.4.4) $$r(t) = f(t)/\bar{F}(t) \quad (t \geq 0)$$

when $\bar{F}(t) > 0$. Noting that

(1.4.5) $$r(t)dt = P\{t < X \leq t+dt \mid X > t\},$$

we can interpret that $r(t)dt$ denotes the conditional probability that the item fails between $(t, t+dt]$ given that the item survives up to time t. Assuming that $F(0) = 0$ (i.e., $\bar{F}(0) = 1$), we have

(1.4.6) $$\bar{F}(t) = \exp[-\int_0^t r(x)dx],$$

and

(1.4.7) $$f(t) = r(t)\exp[-\int_0^t r(x)dx].$$

The reliability function (or distribution) and the probability density can be easily expressed in terms of its failure rate $r(t)$. In particular, $\int_0^t r(x)dx$ is called the cumulative hazard or hazard function.

(1.4.8) Definition If the failure rate $r(t)$ is non-decreasing, then the lifetime distribution $F(t)$ is called Increasing Failure Rate (IFR). If $r(t)$ is non-increasing, then $F(t)$ is called Decreasing Failure Rate (DFR).

It is well-known that the typical failure rate curve can be depicted in Fig. 1.4.1 In general, the failure rate is DFR in the early phase (infant phase), is constant in the middle phase (useful life phase), and is IFR in the final phase (wearout phase). This phenomenon can be found in the human beings as well as the items. Such a curve in Fig. 1.4.1 is called the "Bath-Tub" curve since the curve is quite similar to a bath-tub.

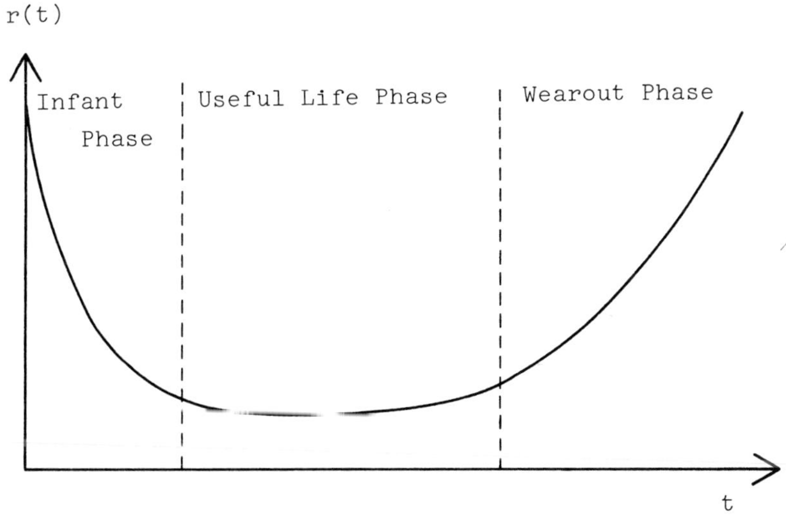

Fig. 1.4.1. A "Bath-Tub" curve of the failure rate r(t).

We are ready to describe typical distributions in reliability theory.

(i) Exponential Distribution

(1.4.9) $F(t) = 1 - e^{-\lambda t} \quad (t \geq 0, \lambda \geq 0),$

(1.4.10) $E[X] = 1/\lambda,$

(1.4.11) $Var(X) = 1/\lambda^2,$

(1.4.12) $r(t) = \lambda.$

The failure rate is constant if and only if $F(t)$ is exponentially distributed. This means that the item which has the exponential distribution never ages. That is, the memoryless property corresponds to the constant failure rate.

(ii) Gamma Distribution

(1.4.13) $f(t) = \lambda e^{-\lambda t}(\lambda t)^{n-1}/\Gamma(n) \quad (t \geq 0, \lambda > 0, n > 0)$

(1.4.14) $E[X] = n/\lambda,$

(1.4.15) $Var(X) = n/\lambda^2,$

where $\Gamma(n) = \int_0^\infty e^{-x}x^{n-1}dx$ is a gamma function of order n. In particular, when n is a positive integer, $\Gamma(n) = (n-1)!$, which implies that the random variable of the gamma distribution is the sum of the independent and identically distributed n random variables with exponential distribution having parameter λ. It is clear that $n = 1$ implies an exponential distribution having parameter λ.

Let us examine the monotone property of the failure rate for the gamma distribution. Noting that

$$(1.4.16) \qquad \bar{F}(t) = \int_t^\infty [\lambda e^{-\lambda x}(\lambda x)^{n-1}/\Gamma(n)]dx$$

and

$$(1.4.17) \qquad [r(t)]^{-1} = \bar{F}(t)/f(t)$$

$$= \int_t^\infty (x/t)^{n-1} e^{-\lambda(x-t)} dx$$

$$= \int_0^\infty (1 + u/t)^{n-1} e^{-\lambda u} du$$

where the change of the variable $u = x - t$ is made, we can show that $r(t)$ is decreasing when $0 < n \leq 1$, and increasing when $n \geq 1$.

When n is a positive integer, we have

$$(1.4.18) \qquad F(t) = \int_0^t [\lambda e^{-\lambda t}(\lambda x)^{n-1}/(n-1)!]dx$$

$$= \sum_{i=n}^{\infty} e^{-\lambda t}(\lambda t)^i/i!$$

$$= 1 - \sum_{i=0}^{n-1} e^{-\lambda t}(\lambda t)^i/i! ,$$

which can be verified by the n iterations of integration by parts. It will be easily verified from the relationship between the Poisson process and the exponential distribution. The failure rate can be expressed by

$$(1.4.19) \qquad r(t) = [\lambda(\lambda t)^{n-1} e^{-\lambda t}/(n-1)!]/\sum_{i=0}^{n-1} e^{-\lambda t}(\lambda t)^i/i! ,$$

which can be calculated from the tables of the Poisson distribution.

(iii) Weibull Distribution

(1.4.20) $F(t) = 1 - \exp[-(\lambda t)^m] \quad (t \geq 0, \lambda > 0, m > 0)$

(1.4.21) $E[X] = (1/\lambda)\Gamma(1+1/m)$,

(1.4.22) $\mathrm{Var}(X) = (1/\lambda^2)\{\Gamma(1+2/m) - \Gamma^2(1+1/m)\}$,

(1.4.23) $r(t) = m\lambda^m t^{m-1}$,

where the parameter λ is called the <u>scale parameter</u> and m the <u>shape parameter</u>. When $m = 1$, the distribution is exponential. Noting that $r(t) = m\lambda^m t^{m-1}$, when $0 < m \leq 1$, the distribution is DFR, and when $m \geq 1$, the distribution is IFR.

The reciprocal of the reliability function of the Weibull distribution is given by

(1.4.24) $1/\bar{F}(t) = \exp[(\lambda t)^m]$.

Taking the logarithms twice yields

(1.4.25) $\log \log 1/\bar{F}(t) = m \log t + m \log \lambda$.

The <u>Weibull probability paper</u> is the functional graph paper with ordinate $\log \log 1/\bar{F}(t)$ and abscissa $\log t$. If we can plot a curve of equation (1.4.25) on the Weibull probability paper, it is natural that the plotted curve is a straight line. The Weibull probability paper can be applied to test the goodness-of-fit of the lifetime data and then the desired parameters can be obtained from the Weibull probability paper.

In practical applications of the Weibull probability paper, we have to specify the plotting positions on the Weibull probability paper for given lifetime data of relatively small size. That is, if n lifetime (to failure) data are given by

$$t_1 < t_2 < \ldots < t_n, \tag{1.4.26}$$

we have to specify the ordinate $F(t_i)$ $(i = 1, 2, \ldots, n)$. Among several proposals for plotting positions, we just present the following three plotting positions:

a) Mean Rank Plotting:

$$F(t_i) = i/(n+1) \quad (i = 1, 2, \ldots, n). \tag{1.4.27}$$

b) Median Rank Plotting: Median rank plotting positions can be obtained from the median of the order statistics, be given by a table, and be well approximated by

$$F(t_i) = (i - 0.3)/(n + 0.4) \quad (i = 1, 2, \ldots, n). \tag{1.4.28}$$

c) Midpoint Rank Plotting:

$$F(t_i) = (i - 0.5)/n \quad (i = 1, 2, \ldots, n). \tag{1.4.29}$$

Applying one of the suitable plottings above, we can plot the Weibull probability paper for the given lifetime data.

If we can draw a straight line through the plotted curve, we can estimate the parameters λ and m from the straight line on the Weibull probability paper. The details of median rank will be discussed in Section 1.5 and the percentile values of the median rank of the ith order statistics for $n = 5\ (1)\ 20$ will be given in Table 1.5.1.

The cumulative hazard of the Weibull distribution is given by

$$(1.4.30) \qquad R(t) = \int_0^t r(x)dx = (\lambda t)^m .$$

Taking the logarithm yields

$$(1.4.31) \qquad \log R(t) = m \log t + m \log \lambda ,$$

which is a linear function of $\log t$ on log-log functional graph paper. The Weibull hazard paper is used to test the goodness-of-fit and to estimate for the lifetime data, in particular, censored data. The readers interested in general probability and hazard plotting should consult books by King (1971) and Nelson (1982).

(iv) Truncated Normal Distribution

The domain of the normal distribution is $(-\infty, \infty)$. If we apply the normal distribution to the lifetime distribution, the domain has to be $[0, \infty)$. That is, the truncated normal distribution is the following probability density:

(1.4.32) $f(t) = (1/a\sqrt{2\pi}\sigma)\exp[-(t-\mu)^2/2\sigma^2]$

$(t \geq 0, -\infty < \mu < \infty, \sigma > 0)$,

where

(1.4.33) $a = \int_0^\infty (1/\sqrt{2\pi}\sigma)\exp[-(x-\mu)^2/2\sigma^2]dx$

is a normalizing constant and is introduced to hold the total probability $\int_0^\infty f(x)dx$ is a unity. For instance, if $\mu = 3\sigma$, $a \cong 0.9987$. If $\mu = 2.5\sigma$, $a \cong 0.9938$. In practice, if $\mu \geq 2.5\sigma$, we assume $a \cong 1$. This distribution has IFR (see the proof by Barlow and Proschan (1975), pp. 76-77).

(v) Log Normal Distribution

The domain of the normal distribution is $(-\infty, \infty)$. If $\log t$ $(t \geq 0)$ is distributed normally, the random variable t is distributed with the following density:

(1.4.34) $f(t) = (1/t\sqrt{2\pi}\sigma)\exp[-(\log t - \mu)^2/2\sigma^2]$

$(t \geq 0, -\infty < \mu < \infty, \sigma > 0)$

which is the probability density of the log normal distribution. The mean and variance are given by

(1.4.35) $E[X] = \exp(\mu+\sigma^2/2)$,

(1.4.36) $Var(X) = \exp(2\mu+2\sigma^2) - \exp(2\mu+\sigma^2)$.

The log normal distribution can be adapted to the

maintenance or repair time distribution. The failure rate curve is increasing (IFR) in the early phase and decreasing (DFR) after that.

1.4.2 Discrete Distributions

In this subsection we are interested in the discrete distributions in reliability theory. Unless otherwise specified, the probability mass function of the discrete random variable is given by $p(k)$ $(k = 1, 2, 3, \ldots)$. The failure rate $r(k)$ $(k = 1, 2, \ldots)$ is defined by

$$(1.4.37) \qquad r(k) = p(k) \Big/ \sum_{j=k}^{\infty} p(j),$$

if $\sum_{j=k}^{\infty} p(j) > 0$. The reliability function and probability mass function can be expressed in terms of its failure rate $r(k)$:

$$(1.4.38) \qquad \bar{F}(k) = \sum_{j=k}^{\infty} p(j) = \prod_{j=1}^{k-1} [1 - r(j)],$$

$$(1.4.39) \qquad p(k) = r(k) \prod_{j=1}^{k-1} [1 - r(j)],$$

where $\prod_{j=1}^{0} \cdot = 1$ for $k = 1$.

(1.4.40) Definition If the failure rate $r(k)$ is non-decreasing in k, then the distribution is IFR. If $r(k)$ is non-increasing, then the distribution is DFR (see the definitions of IFR and DFR for the continuous distribution).

We briefly describe typical discrete distributions in reliability theory.

(i) Geometric Distribution

(1.4.41) $$p(k) = p(1-p)^{k-1} \qquad (k = 1, 2, \ldots;\ 0 < p < 1),$$

(1.4.42) $$r(k) = p(k) \Big/ \sum_{j=k}^{\infty} p(j) = p.$$

The failure rate is constant if and only if $p(k)$ is geometrically distributed. That is, the geometric distribution corresponds to the exponential distribution which is continuous.

(ii) Negative Binomial (or Pascal) Distribution

(1.4.43) $$p(k) = \binom{x-1}{x-r} p^r q^{x-r} = \binom{-r}{x-r} p^r (-q)^{x-r}$$

The negative binomial distribution is IFR when r is a positive integer greater than one as be expected (see Example (1.2.36)). That is, the negative binomial distribution corresponds to the gamma distribution which is continuous.

(iii) Poisson Distribution

(1.4.44) $$p(k) = e^{-\lambda}\lambda^k/k! \qquad (k = 0, 1, 2, \ldots;\ \lambda > 0).$$

The Poisson distribution is IFR.

(iv) Discrete Weibull Distribution

Nakagawa and Osaki (1974) introduced a discrete Weibull distribution corresponding to the continuous one. That is,

$$p(k) = q^{(k-1)^{\beta}} - q^{k^{\beta}} \quad (k=1,2,\ldots;\ \beta>0,\ 0<q<1), \tag{1.4.45}$$

$$r(k) = 1 - q^{k^{\beta}-(k-1)^{\beta}} . \tag{1.4.46}$$

The discrete Weibull distribution is DFR when $0 < \beta \leq 1$ and IFR when $\beta \geq 1$.

1.4.3 Classes of IFR (DFR) Distributions

We return to the continuous time lifetime distributions in general unless otherwise specified.

(1.4.47) Definition A distribution $F(t)$ is called Increasing Failure Rate (IFR) if $F(t)$ satisfies that

$$\bar{F}(x|t) = \bar{F}(t+x)/\bar{F}(t) \tag{1.4.48}$$

is decreasing in $-\infty < t < \infty$ for each $x \geq 0$.

Note that $\bar{F}(x|t) = P\{X > t+x | X > t\}$ is the survival probability that the item is survived up to time $t+x$ given that it was survived up to time t.

(1.4.49) Definition A distribution $F(t)$ is called Decreasing Failure Rate (DFR) if $F(t)$ satisfies that $\bar{F}(x|t)$ is increasing in $t \geq 0$ for each $x \geq 0$.

It is clear that, when the density $f(t)$ exists,

$$(1.4.50) \qquad r(t) = \lim_{x \to 0} (1/x)[1 - \bar{F}(x|t)] \quad (t \geq 0),$$

which corresponds to Definition (1.4.8) in Subsection 1.4.1. For the discrete random variable, if we specify $x = 1$ and $t = k$, then

$$(1.4.51) \qquad \bar{F}(x|t) = \bar{F}(k+1)/\bar{F}(k) = 1 - p(k)/\sum_{j=k}^{\infty} p(j),$$

which corresponds to Definition (1.4.40) in Subsection 1.4.2.

We introduce a broader class of distributions rather than IFR (DFR).

(1.4.52) Definition A distribution $F(t)$ is called <u>Increasing Failure Rate Average</u> (IFRA) if $-(1/t) \log \bar{F}(t)$ is increasing in $t \geq 0$. Similarly, a distribution $F(t)$ is called <u>Decreasing Failure Rate Average</u> (DFRA) if $-(1/t) \log \bar{F}(t)$ is decreasing in $t \geq 0$.

Note that, if the failure rate $r(t)$ exists, then

$$(1.4.53) \qquad (-1/t)\log \bar{F}(t) = (1/t)\int_0^t r(x)dx$$

by which we can understand naming of IFRA (DFRA). It is

clear that the cumulative hazard $R(t) = \int_0^t r(x)dx$ is a convex (concave) function with $R(0) = 0$ when $F(t)$ is IFR (DFR). That is the tangent at t is given by $-(1/t) \log \bar{F}(t)$ which is increasing (decreasing) when $F(t)$ is IFR (DFR). We have the following:

(1.4.54) Theorem If $F(t)$ is IFR (DFR), then $F(t)$ is IFRA (DFRA).

It is also clear that an IFRA (DFRA) distribution $F(t)$ is equivalent to that $\bar{F}(t)^{1/t}$ is decreasing (increasing) on $[0, \infty)$. This fact will be used later.

A few broader classes of distributions will be introduced for replacement problems which will be discussed in Chapter 4. We first introduce the following:

(1.4.55) Definition A distribution is called <u>New Better Than Used</u> (NBU) if

(1.4.56) $$\bar{F}(x+y) \leq \bar{F}(x)\bar{F}(y) \qquad \text{for } x \geq 0,\ y \geq 0.$$

A distribution is called <u>New Worse Than Used</u> (NWU) if

(1.4.57) $$\bar{F}(x+y) \geq \bar{F}(x)\bar{F}(y) \qquad \text{for } x \geq 0,\ y \geq 0.$$

It is obvious that $\bar{F}(x+y) = \bar{F}(x)\bar{F}(y)$ for $x \geq 0$, $y \geq 0$ if and only if $\bar{F}(t) = e^{-\lambda t}$, i.e., the distribution $F(t)$ is exponential. In general, we consider the conditional survival probability

(1.4.58) $$P\{X > x+y \mid X > x\} = \bar{F}(x+y)/\bar{F}(x)$$

is less (greater) than $\bar{F}(y)$ if $F(t)$ is NBU (NWU), by which we can understand naming of NBU (NWU). We are ready to show the following:

(1.4.59) Theorem If $F(t)$ is IFRA (DFRA), then $F(t)$ is NBU (NWU).

Proof. If $F(t)$ is IFRA (DFRA), $\bar{F}(t)^{1/t}$ is decreasing (increasing). Noting that $x \le x+y$, we have $\bar{F}(x)^{1/x} \ge (\le)$ $\bar{F}(x+y)^{1/(x+y)}$ which implies

$$\bar{F}(x) \ge (\le) \bar{F}(x+y)^{x/(x+y)}, \tag{1.4.60}$$

and $\bar{F}(y)^{1/y} \ge (\le) \bar{F}(x+y)^{1/(x+y)}$ which implies

$$\bar{F}(y) \ge (\le) \bar{F}(x+y)^{y/(x+y)}, \tag{1.4.61}$$

if $F(t)$ is IFRA (DFRA). Multiplying each right-hand side and left-hand side in (1.4.60) and (1.4.61), respectively, we have

$$\bar{F}(x)\bar{F}(y) \ge (\le) \bar{F}(x+y)^{(x+y)/(x+y)} = \bar{F}(x+y), \tag{1.4.62}$$

which proves that $F(t)$ is NBU (NWU).

We finally introduce the following:

(1.4.63) Definition A distribution $F(t)$ is called New Better Than Used in Expectation (NBUE) (New Worse Than Used in Expectation (NWUE)) if

(a) $F(t)$ has finite (finite or infinite) mean $1/\mu$,
(b) $\int_t^\infty \bar{F}(x)dx \le (\ge) \bar{F}(t)/\mu$.

If a distribution $F(t)$ is NBUE, then

(1.4.64) $E[X|X > t] = \int_t^\infty \bar{F}(x)dx/\bar{F}(t) \leq (\geq)\ 1/\mu$,

which can be interpreted that the conditional expectation that the item was survived up to time t is less (greater) than or equal to the mean of the new item.

(1.4.65) Theorem If $F(t)$ is NBU (NWU), then $F(t)$ is NBUE (NWUE).

<u>Proof</u>. If $F(t)$ is NBU (NWU), then

(1.4.66) $\bar{F}(x+y) \leq (\geq)\ \bar{F}(x)\bar{F}(y)$.

Integrating from 0 to infinity with respect to y for both sides yields

(1.4.67) $\int_0^\infty \bar{F}(x+y)dy = \int_x^\infty \bar{F}(y)dy \leq (\geq)\ \bar{F}(x) \int_0^\infty \bar{F}(y)dy = \bar{F}(x)/\mu$,

which implies (b) of the definition of NBUE (NWUE). It is obvious to prove that $1/\mu$ is finite if $F(t)$ is NBU since the assumption that $1/\mu$ is infinite implies a contradiction.

1.5 Order Statistics

Let $X_1, X_2, \ldots, X_n$ be a sample of size n from the independent and identically distributed random variables with distribution $F(t)$ $(t \geq 0)$. That is, $X_1, X_2, \ldots, X_n$ are a set of n independent and identical observations. Then the ordered random variable $X_{1:n} \leq X_{2:n} \leq \cdots \leq X_{n:n}$ are called the <u>order statistics</u> from $F(t)$. We can show that

(1.5.1) $$P\{X_{k:n} \leq t\}$$
$$= \sum_{j=k}^{n} \binom{n}{j}[F(t)]^j[\bar{F}(t)]^{n-j}$$
$$= \frac{n!}{(k-1)!(n-k)!} \int_0^{F(t)} x^{k-1}(1-x)^{n-k}dx ,$$

which is an incomplete beta function, and

(1.5.2) $$E[X_{k:n}] = \int_0^{\infty} \sum_{j=0}^{k-1} \binom{n}{j}[F(t)]^j[\bar{F}(t)]^{n-j}dt,$$

which is derived from the formula (1.2.25).

Median for the lifetime distribution is 50 percentile life. The median rank of ith order statistic for a sample of size n is $F(t) \times 100$ percentile such that $P\{X_{i:n} \leq t\} = 0.5$ (50%), which can be calculated from the tables of the incomplete beta functions. Table 1.5.1 shows the values of the median rank of the ith order statistics for $n = 5\ (1)\ 20$. These values can be approximated by the formula (1.4.28). The approximated formula is quite good in practice.

The minimum or maximum of a simple of size n can be easily obtained:

(1.5.3) $$P\{X_{1:n} \leq t\} = 1 - [\bar{F}(t)]^n,$$

(1.5.4) $$P\{X_{n:n} \leq t\} = [F(t)]^n.$$

Table 1.5.1. Percentile values for the ith order statistics for $n = 5\ (1)\ 20$.

i \ n	5	6	7	8	9	10	11	12	13	14	15	16	17	18	19	20
1	13	11	9.5	8.5	7.5	6.5	6	5.5	5	5	4.5	4	4	4	3.5	3.5
2	31	26	23	20	18	16.5	15	13.5	12.5	12	11	10.5	10	9	8.5	8.5
3	50	42	36	32	29	26	24	22	20	18.5	17.5	16.5	15.5	14.5	14	13
4	69	58	50	44	39	36	32	30	28	26	24	23	21	20	19	18
5	87	74	64	56	50	45	41	38	35	33	31	29	27	26	24	23
6		89	77	68	61	55	50	46	43	40	37	35	33	31	29	28
7			91	80	71	64	59	54	50	47	44	41	39	36	35	33
8				92	82	74	68	62	57	53	50	47	44	42	40	38
9					93	84	76	70	65	60	57	53	50	47	45	43
10						93	85	78	72	67	63	59	56	53	50	48
11							94	86	80	74	69	65	62	58	55	52
12								94	87	81	76	71	67	64	60	57
13									95	88	82	77	73	69	65	62
14										95	89	84	79	74	71	67
15											95	90	85	80	76	72
16												96	90	85	81	77
17													96	91	86	82
18														96	91	87
19															96	92
20																97

From the reliability viewpoint, the minimum or maximum distribution of a sample of size n is the distribution to the first failure of n independent <u>series</u> items, or the distribution to the first failure of n independent <u>parallel</u> items.

In particular, if we assume the exponential distribution $F(t) = 1 - e^{-\lambda t}$, then

(1.5.5) $$P\{X_{1:n} \leq t\} = 1 - e^{-n\lambda t},$$

(1.5.6) $$P\{X_{n:n} \leq t\} = (1 - e^{-\lambda t})^n,$$

and

(1.5.7) $$E[X_{1:n}] = 1/n\lambda,$$

(1.5.8) $$E[X_{n:n}] = (1/\lambda)[1 + \frac{1}{2} + \frac{1}{3} + \ldots + \frac{1}{n}] .$$

That is, the minimum distribution is also exponential with parameter $n\lambda$. However, the maximum distribution is no longer exponential.

We cite the redundancy notation from IEEE Transactions on Reliability (see the rear of each Transactions):

k-out-of-n: G The system is good if and only if at least k of its elements are good.

k-out-of-n: F The system is failed if and only if at least k of its elements are failed.

Following the above notation, the distribution to the first failure of a k-out-of-n: G system is $P\{X_{n-k+1:n} \leq t\}$ in

(1.5.1), and the distribution to the first failure of a k-out-of-n: F system is $P\{X_{k:n} \leq t\}$ in (1.5.1). That is, the dual of a k-out-of-n: G system is a k-out-of-n: F system, and vice versa. It is easy to show that

$$(1.5.9) \qquad E[X_{k:n}] = \frac{1}{\lambda}\left(\frac{1}{n} + \frac{1}{n-1} + \ldots + \frac{1}{n-k+1}\right),$$

which can be easily derived by noting the memoryless property of the exponential distribution. Of course, $E[X_{n-k+1:n}]$ is the <u>Mean-Time-To-Failure</u> (MTTF) of a k-out-of-n: G system with independent and identically distributed exponential distribution $F(t) = 1 - e^{-\lambda t}$.

The extreme value theory is concerned with the extreme order statistics $X_{k:n}$, where k remains fixed as $n \to \infty$. In particular, we are very much interested in the limiting distributions for a maximum random variable $X_{n:n}$ or minimum $X_{1:n}$. Specifying the sequences of constants $\{a_n\}$ and $\{b_n\}$ with $a_n > 0$, changing the variables $(X_{n:n} - b_n)/a_n$ and $(X_{1:n} + b_n)/a_n$, and limiting $n \to \infty$, we can obtain all the possible limiting types of distributions which are called the <u>extreme value distributions</u>.

(1.5.10) Theorem If $(X_{n:n} - b_n)/a_n$ has a limiting distribution

$$(1.5.11) \qquad G(y) = \lim_{n\to\infty} P\{(X_{n:n} - b_n)/a_n \leq y\},$$

then $G(y)$ must be one of the following three types of extreme value distributions:

Type I (Exponential type)

$$(1.5.12) \qquad G_1(y) = \exp(-e^{-y}) \qquad (-\infty < y < \infty).$$

Type II (Cauchy)

(1.5.13) $G_2(y) = \exp(-y^{-\alpha})$ $(y \geq 0, \alpha > 0)$.

Type III (Limited)

(1.5.14) $G_3(y) = \begin{cases} \exp[-(-y)^{\alpha}] & (y < 0, \alpha > 0) \\ 1 & (y \geq 0). \end{cases}$

(1.5.15) Theorem If $(X_{1:n} + b_n)/a_n$ has a limiting distribution

(1.5.16) $H(y) = \lim_{n\to\infty} P\{(X_{1:n} + b_n)/a_n \leq y\}$,

then $H(y)$ must be one of the following three types of extreme value distributions:

Type I (Exponential)

(1.5.17) $H_1(y) = 1 - \exp(-e^{-y})$ $(-\infty < y < \infty)$.

Type II (Cauchy)

(1.5.18) $H_2(y) = 1 - \exp[-(-y)^{\alpha}]$ $(y \leq 0, \alpha > 0)$.

Type III (Limited)

(1.5.19) $H_3(y) = 1 - \exp(-y^{\alpha})$ $(y \geq 0, \alpha > 0)$.

Note that $H_3(y)$ is the Weibull distribution discussed in Section 1.4. The distribution $G_1(y)$ and $H_1(y)$ are called the doubly exponential distributions for maximums and minimums, respectively, because of their forms.

Bibliography and Comments

Sections 1.1, 1.2, and 1.3: Many textbooks on probability theory have been published. See Chung (1974), Feller (1968, 1971), and Parzen (1960).

Section 1.4: For the lifetime distributions in reliability theory, see Mann, Shafer and Singpurwalla (1974), Barlow and Proschan (1965, 1975), Bain (1978), and Nelson (1982).

Barlow and Proschan (1975) summarized the notions of aging thoroughly. The extensions have been done and are still done by many researchers.

The Weibull probability plotting and other plotting have been extensively discussed by King (1960). Nelson (1982) also summarized the probability plotting. A variety of probability papers including the Weibull, Weibull hazard and extreme value distributions can be ordered from: Technical and Engineering Aids for Management (TEAM), Box 25, Tamworth, New Hampshere 03886, U.S.A. In Japan, the similar probability papers can be obtained from JUSE (Japan Union of Science and Engineering) and Ostrich Company, Ltd.

Osaki, Kaio and Arita (1981), and Kaio and Osaki (1984) discussed the computer-aided Weibull and Weibull hazard probability papers by using Microcomputer.

Section 1.5: Mann, Shafer and Singpurwalla (1974), Barlow and Proschan (1975), and Bain (1978) discussed extensively the extreme value theory. Gumbel (1964) summarized the extreme value theory and its applications.

[1] L.J. Bain (1978), Statistical Analysis of Reliability and Life-Testing Models: Theory and Methods, Dekker, New York.

[2] R.E. Barlow and F. Proschan (1965), Mathematical Theory of Reliability, Wiley, New York.

[3] R.E. Barlow and F. Proschan (1975), Statistical Theory of Reliability and Life Testing: Probability Models, Holt, Rinehart, and Winston, New York.

[4] K.L. Chung (1974), Elementary Probability Theory with Stochastic Processes, Springer-Verlag, New York.

[5] W. Feller (1968), An Introduction to Probability Theory and Its Applications, Vol. 1, Third Edition, Wiley, New York.

[6] W. Feller (1971), An Introduction to Probability Theory and Its Applications, Vol. 2, Second Edition, Wiley, New York.

[7] N. Kaio and S. Osaki (1984), "The Computer-Aided Weibull Hazard Paper by Microcomputer," International J. Policy and Information, Vol. 8, pp. 65-71.

[8] J.R. King (1971), Probability Charts for Decision Making, Industrial Press, New York.

[9] N.R. Mann, R.E. Shafer and N.D. Singpurwalla (1974), Methods for Statistical Analysis of Reliability and Life Data, Wiley, New York.

[10] T. Nakagawa and S. Osaki (1975), "The Discrete Weibull Distribution," IEEE Trans. Reliability, Vol. R-24, pp. 300-301.

[11] W. Nelson (1982), Applied Data Analysis, Wiley, New York.

[12] S. Osaki, N. Kaio and H. Arita (1981), "The Weibull Probability Papers by Micrcomputer," International J. Policy and Information, Vol. 5, pp. 1-13.

[13] E. Parzen (1960), Modern Probability Theory and Its Applications, Wiley, New York.

CHAPTER 2

STOCHASTIC PROCESSES

2.1 Stochastic Processes

We can observe that the random phenomenon is governed by the probability laws at any time. For instance, we are interested in the random phenomenon assuming 'up' or 'down' state of a system at time $t \geq 0$ in reliability theory. A stochastic process $\{X(t), t \in T\}$ is a family of random variable $X(t)$ which describes the probability law at time $t \in T$. In reliability applications, we restrict ourselves to the non-negative time parameter $t \geq 0$ for the index set T.

If the time parameter T is a countable set $T = \{0, 1, \ldots\}$, the process $\{X(n), n = 0, 1, \ldots\}$ is called a <u>discrete-time stochastic process</u>, and if T is a continuum, the process $\{X(t), t \geq 0\}$ is called a <u>continuous-time stochastic process</u>. For a stochastic process $\{X(t), t \in T\}$, a set of all possible values of $X(t)$ is called a <u>state</u>

space. Throughout this book we restrict ourselves to the discrete state space which can be described by the non-negative integers $i = 0, 1, 2, \ldots$, unless otherwise specified. Of course, we are also interested in a stochastic process $\{X(t), t \in T\}$ with continuous state space. An example of the latter is a Brownian process or diffusion process which is of another interest. However, we never intend to discuss the latter in this book.

A continuous-time stochastic process $\{X(t), t \geq 0\}$ is said to have independent increments if for all $0 \leq t_0 < t_1 < \ldots < t_n$, the variables

$$X(t_1) - X(t_0),\ X(t_2) - X(t_1),\ \ldots,\ X(t_n) - X(t_{n-1})$$

are independent, where the difference $X(t_i) - X(t_{i-1})$ is called the increment. The process $\{X(t), t \geq 0\}$ is said to have stationary increments if $X(t+s) - X(t)$ has the same distribution for all $t \geq 0$. The process $\{X(t), t \geq 0\}$ is said to have stationary independent increments if $X(t_2+s) - X(t_1+s)$ has the same distribution for all $t_2 > t_1 \geq 0$ and $s > 0$. A sample function is a realization of the stochastic process, which will be shown in the following sections.

In the sequel of this chapter, Sections 2.2 and 2.3 are devoted to Poisson processes and renewal processes, respectively, which are continuous-time stochastic processes. Section 2.4 is devoted to Markov chains which are discrete-time stochastic processes. Section 2.5 is devoted to Markov processes which are again continuous-time stochastic processes. In the next chapter, we will discuss Markov renewal processes which are powerful tools for analyzing reliability models throughout this book.

2.2 Poisson Processes

A Poisson process is well-known as a counting process describing the random phenomenon. In reliability theory, when we consider the number of failures or repairs in a finite interval (0, t], we shall consider a counting process $\{N(t), t \geq 0\}$, where the random variable N(t) denotes the total number of failures or repairs in a finite interval (0, t]. A Poisson process has been introduced for describing the random phenomena such as the arriving particles to a counter and the arriving customers in queueing theory.

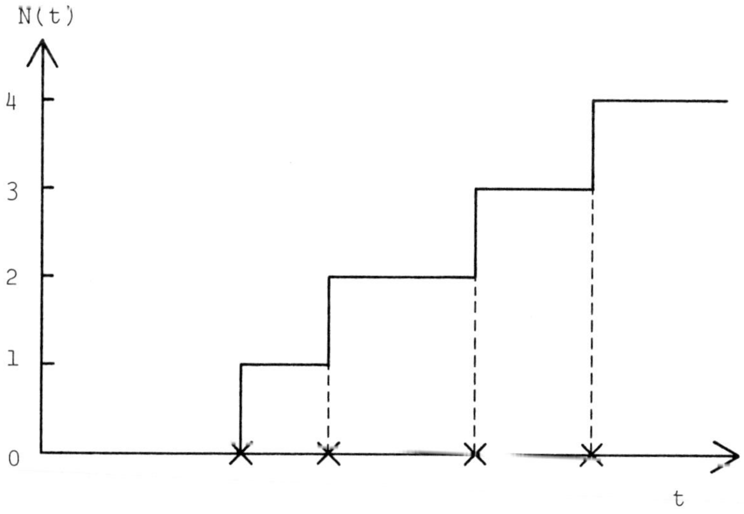

Fig. 2.2.1. A sample function of a counting process $\{N(t), t \geq 0\}$.

Consider the repeated events or occurrences in time, where an 'event' or 'occurrence' is referred to as a failure, a repair, an arriving particle, and an arriving customer, an arriving job, and so on. Fig. 2.2.1 shows a sample function of a counting process $\{N(t), t \geq 0\}$, where the symbol "x" denotes an event or occurrence. Before introducing the definition of a Poisson process, we define $f(h) = o(h)$ (read "small-oh of h") as

$$(2.2.1) \qquad \lim_{h \to 0} f(h)/h = 0.$$

(2.2.2) Example For a small interval $h > 0$, we have

$$(2.2.3) \qquad 1 - e^{-\lambda h} = \lambda h - (\lambda h)^2/2! + (\lambda h)^3/3! - \dots$$

$$= \lambda h + o(h),$$

and

$$(2.2.4) \qquad e^{-\lambda h} = 1 - \lambda h + (\lambda h)^2/2! - \dots$$

$$= 1 - \lambda h + o(h).$$

That is, equations (2.2.3) and (2.2.4) show that the probability that an event takes place for a small interval $h > 0$ is $\lambda h + o(h)$ as $h \to 0$, and the probability that no events take place for a small interval $h > 0$ is $1 - \lambda h + o(h)$ as $h \to 0$, respectively. Note that $P\{X \leq t\} = 1 - e^{-\lambda t}$ and $P\{X > t\} = e^{-\lambda t}$ for the exponential distribution with parameter λ.

We are ready to define a Poisson process as follows:

(2.2.5) Definition A counting process $\{N(t), t \geq 0\}$ is called a Poisson process with parameter $\lambda > 0$ if the following are satisfied:

(i) $N(0) = 0$.
(ii) The process has stationary independent increments.
(iii) $P\{N(h) = 1\} = \lambda h + o(h)$.
(iv) $P\{N(h) \geq 2\} = o(h)$.

Let us define the probability

$$(2.2.6) \qquad P_k(t) = P\{N(t) = k \mid N(0) = 0\} \quad (k = 0, 1, 2, \ldots),$$

for any t, where $P_k(t)$ is the probability mass function for a fixed t. That is,

$$(2.2.7) \qquad \sum_{k=0}^{\infty} P_k(t) = 1$$

for a fixed t, since the total probability is a unity. Noting (ii) in Definition (2.2.5), we have

$$(2.2.8) \qquad P\{N(s+t) - N(s) = k\} = P\{N(t) = k \mid N(0) = 0\} = P_k(t),$$

and noting (ii), (iii), and (iv), we have

$$(2.2.9) \qquad P\{N(t+h) - N(t) = k \mid N(t) = i\} = \begin{cases} 1 - \lambda h + o(h) & (k=i) \\ \lambda h + o(h) & (k=i+1) \\ o(h) & (k>i+1)\ . \end{cases}$$

Using the facts above and Definition (2.2.5), we have

$$P_0(t+h) = P_0(t)[1 - \lambda h + o(h)], \tag{2.2.10}$$

$$\begin{aligned} P_k(t+h) &= P_{k-1}(t)[\lambda h + o(h)] \\ &\quad + P_k(t)[1 - \lambda h + o(h)] \\ &\quad + \sum_{i=2}^{k} P_{k-i}(t)o(h) \quad (k = 1, 2, \ldots). \end{aligned} \tag{2.2.11}$$

Rearranging these terms and assuming $h \to 0$ imply the following set of simultaneous differential equations:

$$P_0'(t) = -\lambda P_0(t), \tag{2.2.12}$$

$$P_k'(t) = \lambda P_{k-1}(t) - \lambda P_k(t) \quad (k = 1, 2, \ldots) \tag{2.2.13}$$

with the initial condition $P_0(0) = P\{N(0) = 0\} = 1$ and $P_k(0) = P\{N(0) = k\} = 0$ $(k = 1, 2, \ldots)$, which can be derived from (i) of Definition (2.2.5). Applying the theory of differential equations, we have

$$P_0(t) = e^{-\lambda t}, \tag{2.2.14}$$

$$P_k(t) = \lambda e^{-\lambda t} \int_0^t e^{\lambda x} P_{k-1}(x)dx \qquad (k = 1, 2, \ldots). \tag{2.2.15}$$

Substituting $P_{k-1}(t)$ into equation (2.2.15) recursively for $k = 1, 2, \ldots$ yields

$$P_k(t) = e^{-\lambda t}(\lambda t)^k/k! \quad (k = 0, 1, 2, \ldots), \tag{2.2.16}$$

which is the probability mass function of the Poisson distribution with parameter λt (see Table 1.2.1). The rigorous verification of equation (2.2.16) requires the techniques of mathematical induction or generating functions.

Let us define the interarrival time or interoccurrence time for each event.

(2.2.17) Definition Let X_1 be the time of the first event or occurrence. In general, let X_n be the time between $(n-1)^{st}$ and n^{th} events. Then $\{X_n, n = 1, 2, \ldots\}$ is called the sequence of interarrival or interoccurrence times.

(2.2.18) Theorem For a Poisson process with parameter λ, the interarrival times X_n $(n = 1, 2, \ldots)$ are independent and identically distributed exponential random variables with mean $1/\lambda$.

Proof. Noting that

(2.2.19) $$P\{X_1 \leq t\}$$

$$= 1 - P\{X_1 > t\} = 1 - P\{N(t) = 0\} = 1 - e^{-\lambda t},$$

we have

(2.2.20)

$$P\{X_2 \leq t | X_1 = s\}$$

$$= 1 - P\{X_2 > t | X_1 = s\}$$

$$= 1 - P\{N(t+s) - N(s) = 0 | X_1 = s\}$$

$$= 1 - P\{N(t+s) - N(s) = 0\} \quad \text{(from independent increments)}$$

$$= 1 - P\{N(t) = 0\} \quad \text{(from stationary increments)}$$

$$= 1 - e^{-\lambda t}.$$

We can recursively show that each interarrival time X_n is independent and identically distributed exponentially. However, we omit the proof in detail.

In Section 2.3 we will introduce a renewal process which is roughly defined as a counting process for which the interarrival times are independent and identically distributed with an arbitrary distribution $F(t)$, say. Then we are ready to give the following:

(2.2.21) Theorem A Poisson process with parameter λ is a renewal process with exponential interarrival distribution $F(t) = 1 - e^{-\lambda t}$.

We are also interested in the arrival or occurrence time of the nth event S_n. Noting that S_n is the sum of independent and identically distributed exponential random variables, i.e.,

(2.2.22) $$S_n = \sum_{i=1}^{n} X_i \qquad (n = 1, 2, \ldots),$$

where we postulate $S_0 = 0$, we have

(2.2.23) $$P\{S_n \leq t\} = \int_0^t [\lambda(\lambda x)^{n-1} e^{-\lambda x}/(n-1)!]dx ,$$

which is the gamma distribution (see Example (1.3.40)).

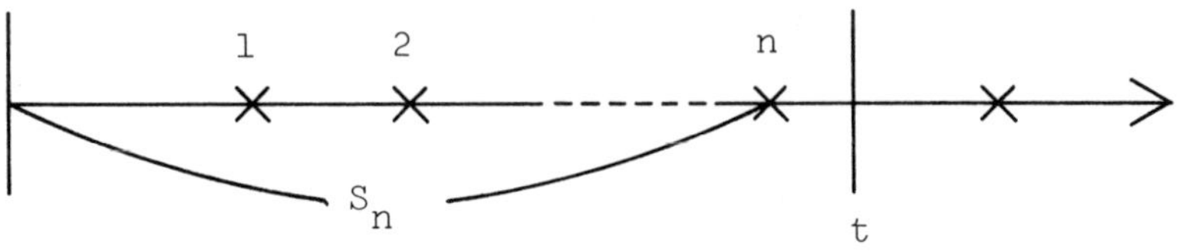

Fig. 2.2.2. A relationship between $N(t)$ and S_n.

As shown in Fig. 2.2.2, we can verify that

(2.2.24) $$S_n \le t \iff N(t) \ge n,$$

i.e., that the arrival time of nth event is less than or equal to t is equivalent to that the total number of events up to time t is greater than or equal to n. Referring to equations (2.2.16), (2.2.23) and (2.2.24), we have the following:

(2.2.25) Theorem For a Poisson process with parameter λ, we have

(2.2.26) $$P\{S_n \le t\} = P\{N(t) \ge n\},$$

or

(2.2.27) $$\int_0^t \frac{\lambda(\lambda x)^{n-1}}{(n-1)!} e^{-\lambda x} dx = \sum_{i=n}^{\infty} e^{-\lambda t} \frac{(\lambda t)^i}{i!} .$$

Identity (2.2.27) can be alternatively verified from analysis of applying n iterations of integration by parts. However, we can easily verify identity (2.2.27) from probabilistic interpretation of the Poisson process.

Consider the conditional distribution of the first arrival time given that there was an event in an interval [0, t]. That is, for $s \leq t$,

(2.2.28) $$P\{X_1 \leq s | N(t)=1\} = P\{N(s)=1,\ N(t)-N(s)=0\}/P\{N(t)=1\} = s/t$$

which is a uniform distribution (see Table 1.2.2). That is, the probability that an event occurs given that there was an event in an interval [0, t] is uniformly distributed over [0, t]. To generalize this fact, we can show the following:

(2.2.29) Theorem The conditional distribution of n arrivals $S_1, S_2, \ldots, S_n$ given that $N(t) = n$ is

(2.2.30) $$P\{S_1 \leq s_1,\ S_2 \leq s_2,\ \ldots,\ S_n \leq s_n | N(t) = n\}$$

$$= n! \int_0^{s_1} \int_{s_1}^{s_2} \cdots \int_{s_{n-1}}^{s_n} \frac{1}{t^n} dx_1 dx_2 \ldots dx_n .$$

That is, the conditional distribution has the same one as the order statistics corresponding to n independent random variables uniformly distributed over the interval [0, t].

The conditional density of n arrivals $S_1, S_2, \ldots, S_n$ given that $N(t) = n$ is

(2.2.31) $$F(t_1, t_2, \ldots, t_n | N(t) = n\} = n!/t^n$$

$$(0 < t_1 < t_2 < \ldots < t_n),$$

which can be interpreted that unordered random variables of n arrivals $S_1, S_2, \ldots, S_n$ given that $N(t) = n$ are independent and identically distributed uniformly over the interval $[0, t]$.

Theorem (2.2.29) can be frequently used for statistical inferences under the assumptions of the exponential lifetime distributions. Of course, Theorem (2.2.29) can be directly applied to several probability models such as the infinite server Poisson queue and the inventory control.

Referring to Fig. 2.2.2, we introduce the following random variables:

(2.2.32) $$\delta_t = t - S_{N(t)},$$

(2.2.33) $$\gamma_t = S_{N(t)+1} - t,$$

(2.2.34) $$\beta_t = \delta_t + \gamma_t$$
$$= S_{N(t)+1} - S_{N(t)} = X_{N(t)},$$

where δ_t is called the <u>excess</u> <u>life</u> or <u>age</u> at time t, γ_t is called the <u>shortage</u> <u>life</u> or <u>residual</u> <u>life</u> at time t, and β_t is called the <u>total</u> <u>life</u> of the $N(t)^{th}$ event. Noting that the Poisson process has stationary independent increments, we have

(2.2.35) $$P\{\delta_t > x\} = P\{\gamma_t > x\} = e^{-\lambda x},$$

and the joint probability that $\delta_t > x$ and $\gamma_t > y$

(2.2.36) $$P\{\delta_t > x, \gamma_t > y\} = e^{-\lambda(x+y)}.$$

Let $\{N_1(t), t \geq 0\}$ and $\{N_2(t), t \geq 0\}$ be two independent Poisson processes with respective parameters λ and μ. Then the pooled process $\{N_1(t) + N_2(t), t \geq 0\}$ is called the <u>superposition of Poisson processes</u>. It is easy to show that

$$(2.2.37) \qquad P\{N_1(t) + N_2(t) = n\}$$

$$= \sum_{k=0}^{n} P\{N_1(t) = k, N_2(t) = n - k\}$$

$$= \sum_{k=0}^{n} e^{-\lambda t}\frac{(\lambda t)^k}{k!} e^{-\mu t}\frac{(\mu t)^{n-k}}{(n-k)!}$$

$$= e^{-(\lambda+\mu)t}\frac{[(\lambda+\mu)t]^n}{n!}$$

which is again a Poisson process with parameter $\lambda + \mu$.

(2.2.38) Theorem The pooled process $\{N_1(t) + N_2(t), t \geq 0\}$ of independent Poisson processes $\{N_1(t), t \geq 0\}$ and $\{N_2(t), t \geq 0\}$ with respective parameters λ and μ is a Poisson process with parameter $\lambda + \mu$.

It is easy to generalize Theorem (2.2.38) with n independent Poisson processes.

Let $\{N(t), t \geq 0\}$ be a Poisson process with parameter λ. Let $\{X(n), n = 1, 2, \ldots\}$ be a Bernoulli process, independent of $N(t)$, with p and $q = 1 - p$ for probabilities of 'success' and 'failure', respectively. The Bernoulli process is associated with the original Poisson process. That is, each event of the original Poisson process can be classified by success or failure which is governed by the Bernoulli process. Such classification is called the <u>decomposition of a Poisson process</u>. Let $\{N_p(t), t \geq 0\}$ and $\{N_q(t), t \geq 0\}$ be the classified processes of success and failure, respectively. Then

(2.2.39) $$P\{N_p(t) = k,\ N_q(t) = n - k\}$$

$$= P\{N_p(t) = k,\ N_q(t) = n - k \mid N(t) = n\}P\{N(t) = n\}$$

$$= \binom{n}{k}p^k q^{n-k} e^{-\lambda t}(\lambda t)^n/n!$$

$$= e^{-p\lambda t}(p\lambda t)^k/k! \cdot e^{-q\lambda t}(q\lambda t)^{n-k}/(n-k)!$$

which implies that $\{N_p(t),\ t \geq 0\}$ and $\{N_q(t),\ t \geq 0\}$ are independent Poisson processes with respective parameters $p\lambda$ and $q\lambda$.

(2.2.40) Theorem The classified process $\{N_p(t),\ t \geq 0\}$ and $\{N_q(t),\ t \geq 0\}$ of success and failure, respectively, are independent Poisson processes with respective parameters $p\lambda$ and $q\lambda$.

It is also easy to generalize Theorem (2.2.40) with n independent classified Poisson processes.

For a Poisson process discussed above, we assume that the process has stationary independent increments. So we should call the <u>stationary</u> or <u>homogeneous</u> <u>Poisson</u> <u>process</u>. Here we simply called the <u>Poisson</u> <u>process</u>. Eliminating the assumption of stationarity, we generalize the nonstationary or nonhomogeneous Poisson process.

(2.2.41) Definition A counting process $\{N(t),\ t \geq 0\}$ is called a <u>nonstationary</u> or <u>nonhomogeneous</u> <u>Poisson</u> <u>process</u> with intensity function $\lambda(t) > 0$ if the following are satisfied:

(i) $N(0) = 0$.

(ii) The process has independent increments.

(iii) $P\{N(t+h) - N(t) = 1 | N(t) = k\} = \lambda(t)h + o(h)$
$(k = 0, 1, 2, \ldots)$.

(iv) $P\{N(t+h) - N(t) \geq 2 | N(t) = k\} = o(h)$
$(k = 0, 1, 2, \ldots)$.

As is similar to the discussions on the Poisson process developed in equations (2.2.6) - (2.2.16), we can show that

$$(2.2.42) \qquad P_k(t) = P\{N(t) = k\}$$
$$= e^{-\Lambda(t)}[\Lambda(t)]^k/k! \qquad (k = 0, 1, 2, \ldots),$$

where

$$(2.2.43) \qquad \Lambda(t) = \int_0^t \lambda(s)ds$$

is called the <u>mean</u> or <u>mean value function</u> since

$$(2.2.44) \qquad E[N(t)] = \Lambda(t).$$

Noting the assumptions of nonstationarity, we have

$$(2.2.45) \qquad P\{N(t+s) - N(s) = k\}$$
$$= e^{-[\Lambda(t+s)-\Lambda(s)]}[\Lambda(t+s) - \Lambda(s)]^k/k!$$
$$(k = 0, 1, 2, \ldots)$$

where

$$(2.2.46) \qquad \Lambda(t+s) - \Lambda(s) = \int_s^{s+t} \lambda(\tau)d\tau .$$

As is just similar to Theorem (2.2.25) of the

stationary Poisson process, for the nonstationary Poisson process with intensity function $\lambda(t) > 0$, we have

(2.2.47) $$P\{S_n \le t\} = P\{N(t) \ge n\}$$

or

(2.2.48) $$\int_0^t \frac{\lambda(t)[\Lambda(t)]^{n-1}}{(n-1)!} e^{-\Lambda(t)} dt = \sum_{i=n}^{\infty} e^{-\Lambda(t)} \frac{[\Lambda(t)]^i}{i!} .$$

The conditional distribution of the first arrival S_1 given that $N(t) = 1$ is

(2.2.49) $$\begin{aligned} P\{S_1 \le s | N(t) = 1\} \\ &= \Lambda(s)e^{-\Lambda(s)}e^{-[\Lambda(t)-\Lambda(s)]}/[\Lambda(t)e^{-\Lambda(t)}] \\ &= \Lambda(s)/\Lambda(t) \\ &= \int_0^s \lambda(\tau)d\tau/\Lambda(t) . \end{aligned}$$

Generalizing this fact, we have the following theorem which corresponds to Theorem (2.2.29) for the nonhomogeneous Poisson process.

(2.2.50) Theorem The conditional distribution of n arrivals $S_1, S_2, \ldots, S_n$ given that $N(t) = n$ is

(2.2.51) $$\begin{aligned} P\{S_1 \le s_1, S_2 \le s_2, \ldots, S_n \le s_n | N(t) = n\} \\ &= n! \int_0^{s_1} \int_{s_1}^{s_2} \cdots \int_{s_{n-1}}^{s_n} \frac{\prod_{i=1}^{n} \lambda(x_i)}{[\Lambda(t)]^n} dx_1 dx_2 \ldots dx_n . \end{aligned}$$

The conditional density of n arrival times S_1, S_2,

..., S_n given that $N(t) = n$ is

$$(2.2.52) \qquad f(t_1, t_2, \ldots, t_n | N(t) = n) = \frac{n!}{[\Lambda(t)]^n} \prod_{i=1}^{n} \lambda(t_i) ,$$

which can be interpreted that unordered random variables of n arrivals S_1, S_2, ..., S_n given that $N(t) = n$ are independent and identically distributed with density

$$(2.2.53) \qquad f(x) = \begin{cases} \lambda(x)/\Lambda(t) & (0 \leq x \leq t) \\ 0 & (\text{otherwise}). \end{cases}$$

In particular, $\Lambda(t) = \lambda t$ implies that $f(x)$ is uniformly distributed over $[0, t]$, which is a special case of the stationary Poisson process given by Theorem (2.2.29).

2.3 Renewal Processes

In the preceding section we have introduced the Poisson process. A renewal process is a generalization of the Poisson process allowing independent and identically distributed arbitrary distributions of the interarrival times. As shown in Theorem (2.2.18), the interarrival times X_n $(n = 1, 2, \ldots)$ are independent and identically distributed exponential random variables for a Poisson process.

(2.3.1) Definition A counting process $\{N(t), t \geq 0\}$ is called a <u>renewal process</u> if the interarrival times are independent and identically distributed with arbitrary distribution $F(t)$.

For a renewal process $\{N(t), t \geq 0\}$, we have introduced interarrival time distributions;

$$(2.3.2) \qquad F(t) = P\{X_k \leq t\} \qquad (k = 1, 2, \ldots)$$

and the distribution of the nth arrival time $S_n = X_1 + X_2 + \cdots + X_n$;

$$(2.3.3) \qquad F^{(n)}(t) = F*F*\ldots*F(t) \qquad (n = 1, 2, \ldots),$$

which is the n-fold Stieltjes convolution with itself. It is convenient for us to define

$$(2.3.4) \quad F^{(n)}(t) = F*F^{(n-1)}(t)$$

$$= \int_0^t F^{(n-1)}(t-x)dF(x) \qquad (n = 1, 2, 3, \ldots)$$

with $F^{(0)}(t) = 1(t)$ (step function) and $F^{(1)}(t) = F(t)$.

Referring to Fig. 2.2.2 for a counting process, we have the following relations:

$$(2.3.5) \qquad S_n \leq t \iff N(t) \geq n$$

which implies

$$(2.3.6) \qquad P\{N(t) \geq n\} = P\{S_n \leq t\} - F^{(n)}(t)$$

and

$$(2.3.7)\qquad P\{N(t) = n\} = P\{N(t) \geq n\} - P\{N(t) \geq n+1\} = F^{(n)}(t) - F^{(n+1)}(t) \qquad (n = 0, 1, 2, \ldots).$$

The mean of the random variable $N(t)$ is called the renewal function $M(t)$ which is the mean number of renewals (or events) up to time t and is given by

$$(2.3.8)\qquad M(t) = E[N(t)] = \sum_{n=1}^{\infty} nP\{N(t) = n\} = \sum_{n=1}^{\infty} P\{N(t) \geq n\} = \sum_{n=1}^{\infty} P\{S_n \leq t\} = \sum_{n=1}^{\infty} F^{(n)}(t).$$

(2.3.9) Example If we assume $F(t) = 1 - e^{-\lambda t}$ for a renewal process, we have

$$(2.3.10)\qquad P\{N(t) = n\} = e^{-\lambda t}(\lambda t)^n/n!$$

and

$$(2.3.11)\qquad M(t) = \lambda t,$$

which were given in Section 2.2 for a Poisson process.

(2.3.12) Example If we assume $F(t) = 1 - (1 + \lambda t)e^{-\lambda t}$, a gamma distribution of order 2 (see Table 1.2.2), we have

$$(2.3.13) \qquad P\{N(t) = n\} = \sum_{i=2n}^{2n+1} e^{-\lambda t}(\lambda t)^i/i! \quad (n=0,1,2,\ldots),$$

and

$$(2.3.14) \qquad M(t) = \lambda t/2 - 1/4 + e^{-2\lambda t}/4 .$$

That is, if we consider a Poisson process with parameter λ, each renewal with gamma interarrival distribution of order 2 takes place, which is a sum of $2n^{th}$ and $(2n+1)^{st}$ terms of the Poisson distribution with parameter λt in (2.3.13).

In general, if we assume that an interarrival distribution $F(t)$ is a gamma distribution of order k (see Table 1.2.2), we have

$$(2.3.15) \qquad P\{N(t) = n\} = \sum_{i=nk}^{nk+k-1} e^{-\lambda t}(\lambda t)^i/i! \ (n=0,1,2,\ldots).$$

We are interested in the asymptotic behaviors of $N(t)/t$ and $M(t)/t$ as $t \to \infty$. Noting that

$$(2.3.16) \qquad S_{N(t)} \leq t \leq S_{N(t)+1},$$

we have

$$(2.3.17) \qquad \frac{S_{N(t)}}{N(t)} \leq \frac{t}{N(t)} \leq \frac{S_{N(t)+1}}{N(t)+1} \cdot \frac{N(t)+1}{N(t)} .$$

From $S_{N(t)}/N(t) = (X_1 + X_2 + \ldots + X_{N(t)})/N(t)$ and the strong law of large numbers, we have

(2.3.18) Theorem With probability 1,

$$(2.3.19) \qquad N(t)/t \to 1/\mu \quad \text{as } t \to \infty ,$$

where $\mu = E[X_1]$, the mean interarrival time.

We can show the following theorem without proof.

(2.3.20) Theorem (Elementary Renewal Theorem)

(2.3.21) $$M(t)/t \to 1/\mu \quad \text{as } t \to \infty,$$

where $1/\mu = 0$ when μ is infinite.

The renewal function $M(t)$ can be rewritten as

(2.3.22) $$\begin{aligned} M(t) &= \sum_{n=1}^{\infty} F^{(n)}(t) \\ &= F(t) + \sum_{n=1}^{\infty} F*F^{(n)}(t) \\ &= F(t) + F*M(t) \\ &= F(t) + \int_0^t M(t-x)dF(x), \end{aligned}$$

which is called the <u>renewal</u> <u>equation</u>. Equation (2.3.22) can be interpreted as an integral equation with known function $F(t)$ and unknown function $M(t)$. Noting the convolution form and introducing the Laplace-Stieltjes transforms

(2.3.23) $$F^*(s) = \int_0^\infty e^{-st} dF(t)$$

and

(2.3.24) $$M^*(s) = \int_0^\infty e^{-st} dM(t),$$

we have

(2.3.25) $$M^*(s) = F^*(s)/[1 - F^*(s)],$$

which implies that the renewal function $M(t)$ is determined uniquely by the interarrival time distribution $F(t)$, and vice versa.

The second moment of $N(t)$ about the origin is

$$(2.3.26) \qquad E[N(t)^2]$$

$$= \sum_{n=1}^{\infty} n^2 P\{N(t)=n\}$$

$$= \sum_{n=1}^{\infty} [2n(n + 1)/2 - n] P\{N(t)=n\}$$

$$= \sum_{n=1}^{\infty} 2 \sum_{k=1}^{n} kP\{N(t)=n\} - \sum_{n=1}^{\infty} \sum_{k=1}^{n} P\{N(t)=n\}$$

$$= \sum_{k=1}^{\infty} 2k \sum_{n=k}^{\infty} P\{N(t)=n\} - \sum_{k=1}^{\infty} \sum_{n=k}^{\infty} P\{N(t)=n\}$$

$$= \sum_{k=1}^{\infty} (2k - 1) P\{N(t) \geq k\}$$

Taking the Laplace-Stieltjes transforms to both sides above, we have

$$(2.3.27) \qquad \int_0^{\infty} e^{-st} dE[N(t)^2]$$

$$= \sum_{k=1}^{\infty} (2k - 1)[F^*(s)]^k$$

$$= 2\left[\frac{F^*(s)}{1 - F^*(s)}\right]^2 + \frac{F^*(s)}{1 - F^*(s)} .$$

Inverting the above, we have

(2.3.28) $$E[N(t)^2] = 2M*M(t) + M(t),$$

and

(2.3.29) $$\mathrm{Var}(N(t)) = 2M*M(t) + M(t) - [M(t)]^2.$$

(2.3.30) Example The Laplace-Stieltjes transform of the geometric distribution is given by

(2.3.31) $$F^*(s) = pe^{-s}/(1 - qe^{-s})$$

(see Table 1.2.1). The Laplace-Stieltjes transform of the renewal function is

(2.3.32) $$M^*(s)$$

$$= \{pe^{-s}/[1 - qe^{-s}]\}/\{1 - pe^{-s}/[1 - qe^{-s}]\}$$

$$= pe^{-s}/(1 - e^{-s})$$

$$= \sum_{k=1}^{\infty} pe^{-ks} .$$

Inverting the Laplace-Stieltjes transform of the renewal function, we have

(2.3.33) $$M(x) = \sum_{k=1}^{x} p = px \quad (x = 1, 2, \ldots),$$

or the probability mass function is $M(x) - M(x-1) = p$ which is constant in x. Note that the discrete-time geometric distribution corresponds to the continuous-time exponential distribution (see Example (2.3.9)).

(2.3.34) Example The Laplace-Stieltjes transform of the negative binomial distribution of order 2 is given by

(2.3.35) $F^*(s) = p^2e^{-2s}/(1 - qe^{-s})^2$

(see Table 1.2.1). The Laplace-Stieltjes transform of the renewal function is

(2.3.36) $M^*(s)$

$$= \frac{p^2e^{-2s}/(1 - qe^{-s})^2}{1 - p^2e^{-2s}/(1 - qe^{-s})^2}$$

$$= \frac{pe^{-s}}{1 - e^{-s}} \cdot \frac{pe^{-s}}{1 - (1 - 2p)e^{-s}}$$

$$= \frac{e^{-s}p}{2}\left[\frac{1}{1 - e^{-s}} - \frac{1}{1 - (1 - 2p)e^{-s}}\right]$$

$$= \sum_{n=1}^{\infty} \frac{p}{2}\{1 - (1 - 2p)^n\}e^{-s(n+1)} ,$$

which implies

(2.3.37) $m(x) = \frac{p}{2}\{1 - (1 - 2p)^{x-1}\} \quad (x = 2, 3, \ldots)$

and

(2.3.38) $M(x)$

$$= \sum_{j=2}^{x} m(j)$$

$$= px/2 - 1/4 + (1 - 2p)^x/4 \qquad (x = 2, 3, \ldots) .$$

Let μ , σ^2, and μ_3 be the mean, the variance, and the third moment about the origin of the interarrival time distribution $F(t)$, respectively. Then we can expand $F^*(s)$ with respect to s:

(2.3.39) $F^*(s) = 1 - \mu s + \frac{1}{2}(\mu^2 + \sigma^2)s^2 - \frac{1}{3!}\mu_3 s^3 + o(s^3)$.

Substituting $F^*(s)$ into (2.3.25) and (2.3.27), rearranging them in terms of s, and applying Tauberian Theorem (see Appendix), we have the following asymptotic results:

$$(2.3.40) \qquad M(t) = \frac{t}{\mu} + \left(1 + \frac{\sigma^2}{2\mu^2}\right) + o(1) ,$$

$$(2.3.41) \qquad Var(N(t)) = \frac{\sigma^2 t}{\mu^3} + \left(\frac{1}{12} + \frac{5\sigma^4}{4\mu^4} - \frac{2\mu_3}{3\mu^3}\right) + o(1) .$$

Barlow and Proschan (1975) showed the following bounds for $M(t)$: If $F(t)$ is NBUE, then

$$(2.3.42) \qquad \frac{t}{\mu} - 1 \le M(t) \le \frac{t}{\mu} \quad (t \ge 0) .$$

Using the asymptotic results in (2.3.40) and (2.3.41), we have the following theorem which corresponds to the central limit theorem for the renewal process:

(2.3.43) Theorem Let μ and σ^2 be the finite mean and variance of the interarrival time distribution $F(t)$. Then

$$(2.3.44) \qquad P\left\{\frac{N(t) - t/\mu}{\sigma\sqrt{t/\mu^3}} < y\right\} \to \frac{1}{\sqrt{2\pi}}\int_{-\infty}^{y} e^{-x^2/2}dx \quad (t \to \infty) .$$

That is, $N(t)$ is asymptotically normal with mean t/μ and variance $t\sigma^2/\mu^3$.

A non-negative random variable X (or a distribution $F(t)$) is said to be <u>lattice</u> or <u>arithmetic</u> if there exists $\delta > 0$ such that $\sum_{n=0}^{\infty} P\{X = n\delta\} = 1$, where δ is called the period of X. The following theorem is well-known as Blackwell's theorem.

(2.3.45) Theorem (i) If $F(t)$ is not lattice, then

(2.3.46) $$M(t+\tau) - M(t) \to \tau/\mu \quad \text{as} \quad t \to \infty$$

for all $\tau \geq 0$. (ii) If $F(t)$ is lattice with period δ, then

(2.3.47) $$M(n\delta) \to \delta/\mu \quad \text{as} \quad n \to \infty.$$

Let $h(t)$ be an arbitrary and bounded function defined on $[0, \infty)$. Let $\underline{m}_k(a)$ and $\bar{m}_k(a)$ be the infimum and superimum over the interval $(k-1)a \leq t \leq ka$ for any $a \geq 0$. Then $h(t)$ is said to be directly Riemann integrable if $\sum_{k=1}^{\infty} \underline{m}_k(a)$ and $\sum_{k=1}^{\infty} \bar{m}_k(a)$ are finite for all $a \geq 0$ and converges the same value as $a \to 0$. We just cite a sufficient condition for $h(t)$ to be directly Riemann integrable:

(i) $h(t) \geq 0$ for all $t \geq 0$.
(ii) $h(t)$ is non-increasing.
(iii) $\int_0^\infty h(t)dt < \infty$.

We are now ready to show the key renewal theorem (without proof) which will be of great use in reliability theory:

(2.3.48) Theorem If $F(t)$ is not lattice and $h(t)$ is directly Riemann integrable, then

(2.3.49) $$\lim_{t\to\infty} \int_0^t h(t-x)dM(x) = \frac{1}{\mu}\int_0^\infty h(t)dt.$$

We shall show how to apply the key renewal theorem in practice. We can generalize the renewal equation to

$$(2.3.50) \qquad g(t) = h(t) + \int_0^t g(t-x)dF(x) \qquad (t \geq 0),$$

where $h(t)$ and $F(t)$ are known and $g(t)$ is unknown in the integral equation (2.3.50). The integral equation (2.3.50) is called the <u>renewal-type equation</u> and is given in terms of the renewal function $M(t)$:

$$(2.3.51) \qquad g(t) = h(t) + \int_0^t h(t-x)dM(x).$$

It is quite easy to derive (2.3.51) from (2.3.50) since the Laplace-Stieltjes transform of $g(t)$ is given by

$$(2.3.52) \qquad \begin{aligned} g^*(s) &= h^*(s)/[1 - F^*(s)] \\ &= h^*(s) + h^*(s)\, F^*(s)/[1 - F^*(s)], \end{aligned}$$

where $h^*(s)$ is the Laplace-Stieltjes transform of $h(t)$. Inverting the Laplace-Stieltjes transforms in (2.3.52) implies equation (2.3.51). Once the renewal-type equation is given and the suitable conditions are satisfied (e.g., $h(t)$ is directly Riemann integrable), we can apply the key renewal theorem to (2.3.51) which implies the asymptotic behavior of $g(t)$ as $t \to \infty$.

A direct application of these results is the residual life distribution $P\{\gamma_t \leq x\}$ (see Section 2.2). Conditioning that the first renewal takes place at $X_1 = y$, we have

$$(2.3.53) \qquad P\{\gamma_t > x | X_1 = y\} = \begin{cases} P\{\gamma_{t-y} > x\} & (y \leq t) \\ 0 & (t < y \leq t+x) \\ 1 & (y > t+x) \end{cases}$$

and

$$(2.3.54) \qquad P\{\gamma_t > x\} = \int_0^\infty P\{\gamma_t > x | X_1 = y\} dF(y).$$

Substituting (2.3.54) into (2.3.53), we have

$$(2.3.55) \qquad P\{\gamma_t > x\} = 1 - F(t+x) + \int_0^t P\{\gamma_{t-y} > x\} dF(y)$$

$$= h(t) + \int_0^t h(t-y) dM(y),$$

since we put $h(t) = 1 - F(t+x)$ and apply (2.3.51) for (2.3.55). Applying the key renewal theorem to (2.3.55) under the suitable conditions, we have

$$(2.3.56) \qquad \lim_{t\to\infty} P\{\gamma_t > x\} = \frac{1}{\mu}\int_0^\infty [1 - F(t+x)] dt$$

$$= \frac{1}{\mu}\int_x^\infty [1 - F(y)] dy.$$

Then we have the following:

(2.3.57) Theorem The residual lifetime distribution is

$$(2.3.58) \qquad P\{\gamma_t \le x\} = F(t+x) - \int_0^t [1 - F(t+x-y)] dM(y)$$

and its asymptotic distribution is

$$(2.3.59) \qquad \lim_{t\to\infty} P\{\gamma_t \le x\} = \frac{1}{\mu}\int_0^x [1 - F(y)] dy$$

if $F(t)$ is not lattice.

For the excess life or age distribution $P\{\delta_t \le x\}$, noting that $\delta_t > x$ is equivalent to $\gamma_{t-x} > x$, we have

(2.3.60) Theorem The excess life or age distribution is

$$(2.3.61) \qquad P\{\delta_t \leq x\} = \begin{cases} F(t) - \int_0^{t-x}[1 - F(t-y)]dM(y) & (x \leq t) \\ 1 & (x > t), \end{cases}$$

and its asymptotic distribution is

$$(2.3.62) \qquad \lim_{t\to\infty} P\{\delta_t \leq x\} = \frac{1}{\mu}\int_0^x [1 - F(y)]dy$$

if $F(t)$ is not lattice.

It is quite interesting that both the asymptotic residual life and excess life distributions are given by the same distribution

$$(2.3.63) \qquad F_e(t) = \frac{1}{\mu}\int_0^t [1 - F(y)]dy$$

which is called the <u>equilibrium distribution</u> of $F(t)$. The equilibrium distribution $F_e(t)$ will be used for a stationary renewal process for the later discussion.

In the preceding discussions on the renewal processes, we have assumed that all the interarrival distributions are identical. However, we can generalize that only the first interarrival time distribution $G(t)$, say, is different from the successive interarrival time distributions $F(t)$.

(2.3.64) Definition A counting process $\{N_D(t), t \geq 0\}$ is called the <u>delayed renewal process</u> if the interarrival times are independent and the first interarrival time is distributed with $G(t)$ and the successive interarrival times are distributed identically with $F(t)$.

It is easy to show that

$$(2.3.65) \qquad P\{N_D(t) = k\} = \begin{cases} 1 - G(t) & (k = 0) \\ G*F^{(k-1)}(t) - G*F^{(k)}(t) & \\ & (k = 1, 2, \ldots), \end{cases}$$

and

$$(2.3.66) \qquad M_D(t) = E[N_D(t)] = \sum_{k=0}^{\infty} G*F^{(k)}(t).$$

The renewal equation is given by

$$(2.3.67) \qquad M_D^*(s) = G^*(s)/[1 - F^*(s)]$$

where $G^*(s)$ is the Laplace-Stieltjes transform of $G(t)$.

The following theorem can be easily verified from the preceding results of the renewal processes. In particular, the asymptotic results are just the same since the first interarrival time distribution $G(t)$ has no contribution to the asymptotic results:

(2.3.69) Theorem For a delayed renewal process $\{N_D(t), t \geq 0\}$:

(i) With probability 1,

$$(2.3.70) \qquad N_D(t)/t \to 1/\mu \quad \text{as} \quad t \to \infty.$$

(ii) (Elementary renewal theorem)

$$(2.3.71) \qquad M_D(t)/t \to 1/\mu \quad \text{as} \quad t \to \infty.$$

(iii) (Blackwell's theorem) If $F(t)$ is not lattice,

(2.3.72) $M_D(t+\tau) - M_D(t) \to \tau/\mu$ as $t \to \infty$,

and if $G(t)$ and $F(t)$ are lattice with period δ,

(2.3.73) $M_D(n\delta) \to \delta/\mu$ as $t \to \infty$.

(iv)

(2.3.74) $P\{\gamma_t \leq x\} = G(t+x) - \int_0^t [1 - F(t+x-y)]dM_D(y)$.

(v) If $F(t)$ is not lattice,

(2.3.75) $\lim_{t\to\infty} P\{\gamma_t \leq x\} = \frac{1}{\mu}\int_0^x [1 - F(y)]dy$.

(vi)

(2.3.76) $P\{\delta_t \leq x\}$

$$= \begin{cases} G(t) - \int_0^{t-x} [1 - F(t-y)]dM_D(y) & (x \leq t) \\ 1 & (x > t). \end{cases}$$

(vii) If $F(t)$ is not lattice,

(2.3.77) $\lim_{t\to\infty} P\{\delta_t \leq x\} = \frac{1}{\mu}\int_0^x [1 - F(y)]dy$.

As a special case of a delayed renewal process, we assume the first interarrival time distribution:

(2.3.78) $G(t) = F_e(t) = \frac{1}{\mu}\int_0^t [1 - F(y)]dy$.

Such a renewal process can be obtained from the renewal process if we assume the starting point (time 0) sufficiently far from the time origin of the underlying renewal process. Then the first interarrival time distribution is given by (2.3.79).

(2.3.79) Definition A counting process $\{N_S(t), t \geq 0\}$ is called the <u>stationary renewal process</u> if the interarrival times are independent and the first interarrival time is distributed with $F_e(t)$ in (2.3.78) and the successive interarrival times are distributed identically with $F(t)$.

The Laplace-Stieltjes transform of the renewal function $M_S(t) = E[N_S(t)]$ is given by

$$(2.3.80) \qquad M_S^*(s) = F_e^*(s)/[1 - F^*(s)] = 1/\mu s ,$$

since $F_e^*(s) = [1 - F^*(s)]/(\mu s)$. Inverting the Laplace-Stieltjes transform of (2.3.80) implies

$$(2.3.81) \qquad M_S(t) = t/\mu$$

which is equal to that of the Poisson process. Substituting (2.3.81) into (2.3.74) and (2.3.76), and noting that $G(t) = F_e(t)$, we have

$$(2.3.82) \qquad P\{\gamma_t \leq x\} = \frac{1}{\mu}\int_0^x [1 - F(y)]dy$$

$$(2.3.83) \qquad P\{\gamma_t \leq x\} = \begin{cases} \frac{1}{\mu}\int_0^x [1 - F(y)]dy & (x \leq t) \\ 1 & (x > t) \end{cases}$$

for all t. Summarizing the results above, we have the following:

(2.3.84) Theorem For a stationary renewal process $\{N_S(t), t \geq 0\}$:

(i)

$$(2.3.85) \qquad M_S(t) = E[N_S(t)] = t/\mu.$$

(ii)

$$(2.3.86) \qquad P\{\gamma_t \leq x\} = \frac{1}{\mu}\int_0^x [1 - F(y)]dy$$

for all $t \geq 0$.

(iii) For all $t, s > 0$,

$$(2.3.87) \qquad P\{N_S(t+s) - N_S(s) = k\} = P\{N_S(t) = k\}$$

i.e., the process $\{N_S(t), t \geq 0\}$ has stationary independent increments.

It is easy to obtain the stationary renewal process from the ordinary or delayed renewal process if we assume the starting point sufficiently far from the time origin and introduce a new process with such a new starting point. For such a new process, the first interarrival time is given by

$$(2.3.88) \qquad P\{\gamma_t \leq x\} = F_e(x) = \frac{1}{\mu}\int_0^x [1 - F(y)]dy$$

from (2.3.59) in Theorem (2.3.57) or (2.3.77) in Theorem (2.3.69).

We conclude with compound processes associated with renewal processes in this section. Consider a renewal process $\{N(t), t \geq 0\}$ and its associated (or the secondary) random variable Y_k $(k = 1, 2, \ldots)$ generated by each renewal.

(2.3.89) Definition For a renewal process $\{N(t), t \geq 0\}$, its associated random variable Y_k $(k = 1, 2, \ldots)$ generated by the kth renewal is assumed to be independent of the original renewal process and other Y_j $(j \neq k)$, and distributed identically with $H(x) = P\{Y_k \leq x\}$. Then, $\{W(t), t \geq 0\}$ is called the compound process or renewal reward process, where

$$(2.3.90) \qquad W(t) = \sum_{k=1}^{N(t)} Y_k.$$

(2.3.91) Definition For a compound or renewal reward process $\{W(t), t \geq 0\}$, if the underlying renewal process is a Poisson process, then $\{W(t), t \geq 0\}$ is called the compound Poisson process.

It is generally true that the state space of the secondary random variable is $(-\infty, \infty)$. However, if we assume the non-negative secondary random variables for a compound process, such a process $\{W(t), t \geq 0\}$ is called the cumulative process since the secondary process is additive.

First, we give the well-known results on the compound Poisson process.

(2.3.92) Theorem For a compound Poisson process $\{W(t), t \geq 0\}$ with parameter λ of the underlying Poisson process, the characteristic function $\phi_W(u)$ of the random variable $W(t)$ is given by

$$(2.3.93) \qquad \phi_W(u) = \exp\{\lambda t[\phi_Y(u) - 1]\},$$

where $\phi_Y(u)$ is the characteristic function of the random variable Y. Differentiating $\phi_W(u)$ with respect to u implies

$$(2.3.94) \quad E[W(t)] = \lambda t E[Y],$$

and

$$(2.3.95) \quad Var(W(t)) = \lambda t E[Y^2].$$

Secondary, we discuss the general compound process or renewal reward process. In this case, we just give the following asymptotic result without proof:

(2.3.96) Theorem For a compound process $\{W(t), t \geq 0\}$, if $\mu = \int_0^\infty t dF(t) < \infty$ and $E[Y] = \int_0^\infty y dG(y) < \infty$, then

(i) With probability 1,

$$(2.3.97) \quad \frac{W(t)}{t} \to \frac{E[Y]}{\mu} \quad \text{as} \quad t \to \infty .$$

(ii)

$$(2.3.98) \quad \frac{E[W(t)]}{t} \to \frac{E[Y]}{\mu} \quad \text{as} \quad t \to \infty .$$

In reliability application we encounter a model which repeats the same stochastic behavior called a <u>cycle</u>. That is, a cycle is completed whenever each renewal takes place for a compound process or renewal reward process. Then, the expected reward per unit time in the steady-state is the ratio of the expected reward of each cycle to the expected duration of each cycle, which is given in (2.3.99).

(2.3.99) Example An age replacement model (which will be discussed in Section 4.2) is a renewal reward process. That is, the underlying renewal process is a truncated renewal process with interarrival time distribution $F(t)$. If an item fails within T, the item is replaced for a spare with constant cost c_1 of corrective maintenance. Otherwise, if the item does not fail up to time T, the item is replaced for a spare with constant cost c_2 of preventive maintenance, where we assume $c_1 > c_2$. A cycle terminates whenever the item fails within T or survives up to time T, whichever occurs first. The cycle repeats with itself again and again. The mean duration for a cycle is given by

$$(2.3.100) \qquad \int_0^T t\,dF(t) + T\bar{F}(T) = \int_0^T \bar{F}(t)dt,$$

and the expected cost for a cycle is $c_1F(T) + c_2\bar{F}(T)$. Applying Theorem (2.3.96), we have the following expected cost per unit time in the steady-state:

$$(2.3.101) \qquad \lim_{t\to\infty} E[W(t)]/t = [c_1F(T) + c_2\bar{F}(T)]/\int_0^T \bar{F}(t)dt.$$

2.4 Markov Chains

Consider a discrete-time stochastic process $\{X(n), n = 0, 1, 2, ...\}$ with state space $i = 0, 1, 2, ...$, unless otherwise specified. That is, $X(n) = i$ denotes that the process is in state i $(i = 0, 1, 2, ...)$ at time n $(n = 0, 1, 2, ...)$. In this section we consider such a discrete-time stochastic process called the Markov chain. A Markov chain plays an important role for studying reliability models as well as elementary stochastic processes. In particular, we shall discuss a Markov renewal process, which is a direct extension of both a Markov chain and a renewal process, in the following chapter.

(2.4.1) Definition For a discrete-time stochastic process $\{X(n),\ n = 0, 1, 2, \ldots\}$, if

(2.4.2)

$$P\{X(n+1)=j \mid X(0)=i_0,\ X(1)=i_1,\ \ldots,\ X(n-1)=i_{n-1},\ X(n)=i\}$$

$$= P\{X(n+1)=j \mid X(n)=i\}$$

$$= p_{ij}$$

for all $i_0, i_1, \ldots, i_{n-1}$, i, j and n, then the process is called a Markov chain and p_{ij} is called a transition probability. In particular, if we assume that p_{ij} in (2.4.2) is independent of time n, the process is called a Markov chain with stationary transition probabilities.

A matrix $\mathbf{P} = [p_{ij}]$ is called a transition probability matrix, where

$$p_{ij} \geq 0, \quad \sum_{j=0}^{\infty} p_{ij} = 1 \quad (i, j = 0, 1, 2, \ldots). \tag{2.4.3}$$

Consider the n-step transition probability

$$p_{ij}^{n} = P\{X(n+m)=j \mid X(m)=i\}, \tag{2.4.4}$$

where we postulate

$$p_{ij}^{0} = 0 \ (i \neq j), \quad p_{ii}^{0} = 1. \tag{2.4.5}$$

Then the n-step transition probability p_{ij}^{n} can be

calculatied by summing over all the intermediate state k at time r and moving to j from k at the remaining time n-r. That is,

$$(2.4.6) \qquad p_{ij}^{n} = \sum_{k=0}^{\infty} p_{ik}^{r} p_{kj}^{(n-r)},$$

which is called the Chapman-Kolmogorov equation. In a matrix form, we have

$$(2.4.7) \qquad \mathbf{P}^{(n)} = \mathbf{P}^{(r)}\ \mathbf{P}^{(n-r)},$$

where $\mathbf{P}^{(n)} = [p_{ij}^{n}]$, and

$$(2.4.8) \qquad \mathbf{P}^{(n)} = \mathbf{P}^{(1)}\ \mathbf{P}^{(n-1)} = \mathbf{P}\ \mathbf{P}^{(n-1)} = \ldots = \mathbf{P}^{n}.$$

That is, the n-step transition probability can be calculated by the nth power of matrix **P**. Rewriting (2.4.7) by using the above fact, we have

$$(2.4.9) \qquad \mathbf{P}^{n} = \mathbf{P}^{r}\ \mathbf{P}^{n-r},$$

which is merely the matrix product.

Introducing an initial distribution vector

$$(2.4.10) \qquad \underline{\pi}(0) = [\pi_0(0),\ \pi_1(0),\ \ldots]\ ,$$

such that

$$(2.4.11) \qquad \pi_j(0) \geq 0 \quad (j = 0, 1, 2, \ldots)\ , \qquad \sum_{j=0}^{\infty} \pi_j(0) = 1,$$

we can calculate the n-step distribution vector

(2.4.12) $\quad \underline{\pi}(n) = [\pi_0(n), \pi_1(n), \ldots]$,
$\quad = \underline{\pi}(0)\mathbf{P}^n \qquad (n = 0, 1, 2, \ldots),$

where $\mathbf{P}^0 = \mathbf{I}$, an identity matrix.

We call that state j can be reached from state i, and write it $i \to j$, if there exists an integer $n \geq 0$ such that $p_{ij}^n > 0$. If $i \to j$ and $j \to i$, that is, there exist integers $m \geq 0$ and $n \geq 0$ such that $p_{ij}^m > 0$ and $p_{ij}^n > 0$, then we call states i and j communicate, and write it $i \leftrightarrow j$. By using communication relation, we can classify all the states of a Markov chain into some equivalence classes.

(2.4.13) Theorem Communication is an equivalence relation. That is,

(i) $i \leftrightarrow i$.
(ii) If $i \leftrightarrow j$, then $j \leftrightarrow i$.
(iii) If $i \leftrightarrow j$ and $j \leftrightarrow k$, then $i \leftrightarrow k$.

If there exists only a single communication class for a Markov chain, we call that the Markov chain is irreducible. That is, for an irreducible Markov chain, all the states communicates each other.

We call that state i has period $d(i)$ if $d(i)$ is the greatest common divisor of $n \geq 1$ such that $p_{ii}^n > 0$. If $d(i) = 1$, then state i is aperiodic, and if $d(i) > 1$, then state i is periodic. We show the following theorem without proof.

(2.4.14) Theorem If $i \leftrightarrow j$, then $d(i) = d(j)$.

Let us define the following probability

(2.4.15) $$f_{ij}^{n} = P\{X(n)=j,\ X(k)\neq j,\ k=1,2,\ldots,n-1 \mid X(0)=i\}$$

for any i, j, which is the first passage probability from state i to state j with n steps, where we postulate $f_{ij}^{0} = 0$ and $f_{ij}^{1} = p_{ij}$. Noting that f_{ij}^{n} is the probability mass function for fixed states i and j, we define

(2.4.16) $$f_{ij} = \sum_{n=1}^{\infty} f_{ij}^{n}$$

which is the total probability from state i to state j.

(2.4.17) Definition If $f_{ii} = 1$, then state i is said to be <u>recurrent</u>. If $f_{ii} < 1$, then state i is <u>transient</u>.

We show the following necessary and sufficient conditions to identify whether state i is recurrent or transient without proof:

(2.4.18) Theorem State i is recurrent if and only if

(2.4.19) $$\sum_{n=1}^{\infty} p_{ii}^{n} = \infty .$$

State i is transient if and only if

(2.4.20) $$\sum_{n=1}^{\infty} p_{ii}^{n} < \infty .$$

It is easy to understand that for any recurrent state i the process revisits to state i infinitely often, which implies equation (2.4.19).

(2.4.21) Theorem If state i is recurrent and $i \leftrightarrow j$, then state j is recurrent and $f_{ij} = 1$.

(2.4.22) Corollary An irreducible Markov chain is either recurrent or transient.

We call that state j is <u>absorbing</u> if j forms an equivalence class by itself. It is easy to verify that state j is absorbing if and only if $p_{jj} = 1$. It is sometimes called an <u>absorbing Markov chain</u> if there is at least an absorbing state.

(2.4.23) Theorem In a Markov chain, all the recurrent states can be classified into some recurrent classes C_1, C_2, ..., and the remaining states which are transient.

Let C_1, C_2, ... be all the recurrent classes and T be a set of all the remaining transient states of a Markov chain. Then all sets C_1, C_2, ..., T are disjoint and exhaustive. Relabeling all the states suitably, we can rewrite

$$\textbf{P} = \begin{array}{c} C_1 \\ C_2 \\ \vdots \\ \\ T \end{array} \left[\begin{array}{cccc|c} \textbf{P}_1 & & \cdots & & \textbf{0} \\ & \textbf{P}_2 & \cdots & & \textbf{0} \\ & & \cdots & & \\ & \cdots & & & \\ \hline \textbf{R}_1 & \textbf{R}_2 & \cdots & & \textbf{Q} \end{array} \right] \qquad (2.4.24)$$

where the submatrices $\textbf{P}_1$, $\textbf{P}_2$, ... are the transition probability matrices for respective recurrent classes C_1, C_2, ..., $\textbf{Q}$ is a square matrix describing the transitions from all the transient states for T, and $\textbf{R}_1$, $\textbf{R}_2$, ... are (not necessarily square) matrices describing the transitions from all the transient states to the corresponding recurrent classes C_1, C_2, ...

We have classified all the states into recurrent or transient states by $f_{jj} = 1$ or $f_{jj} < 1$, respectively. We have to further classify all the recurrent states into positive or null recurrent states by $\mu_j < \infty$ or $\mu_j = \infty$, respectively, where

$$(2.4.25) \qquad \mu_j = \sum_{n=1}^{\infty} n f_{jj}^n$$

is the mean recurrence time for state j.

Let us consider the limiting behavior of p_{jj}^n as $n \to \infty$. Focussing on a specified state j, the Markov chain starts from state j, visits to itself with probability mass function f_{jj}^n $(n = 1, 2, \ldots)$, and so on. That is, the Markov chain is regarded as a discrete-time renewal process for a specified state j. Applying Blackwell's Theorem (2.3.45), we have the following:

(2.4.26) Theorem If state j is recurrent and aperiodic, then

$$(2.4.27) \qquad p_{jj}^n \to 1/\mu_j$$

as $n \to \infty$. And if state j is recurrent and periodic with period $d(j)$, then

$$(2.4.28) \qquad p_{jj}^{nd(j)} \to d(j)/\mu_j \quad \text{as} \quad n \to \infty ,$$

where we interpret $1/\mu_j = 0$ if $\mu_j = \infty$ (i.e., if state j is null recurrent).

(2.4.29) Corollary If state j is transient, then

$$(2.4.30) \qquad P^n_{jj} \to 0 \qquad \text{as } n \to \infty.$$

Let us consider the limiting behavior of p^n_{ij} as $n \to \infty$ in general. Again we focus on specified states i and j, where state i is a starting state. The Markov chain starts from state i, visits to state j with probability mass function f^n_{ij}, and infinitely often revisits to state j with probability mass function f^n_{jj}, which is regarded as a delayed renewal process. Noting that f_{ij} denotes the eventual probability from state i to state j, we have the following:

(2.4.31) Theorem If state j is recurrent and aperiodic, then

$$(2.4.32) \qquad p^n_{ij} \to f_{ij}/\mu_j$$

as $n \to \infty$, where we interpret $1/\mu_j = 0$ if $\mu_j = \infty$.

We are now ready to derive the limiting probabilities for an irreducible Markov chain which is positive recurrent and aperiodic (such a Markov chain is sometimes called an <u>ergodic Markov chain</u>).

(2.4.33) Theorem If an irreducible Markov chain is positive recuurent and aperiodic, there exists the limiting probability

$$(2.4.34) \qquad \lim_{n\to\infty} p^n_{ij} = \pi_j > 0 \quad (i, j = 0, 1, 2, \ldots),$$

which is independent of the initial state i, where $\{\pi_j, j = 0, 1, 2, \ldots\}$ is a unique and positive solution to

$$(2.4.35) \qquad \pi_j = \sum_{i=0}^{\infty} \pi_i \, p_{ij} \qquad (j = 0, 1, 2, \ldots) ,$$

$$(2.4.36) \qquad \sum_{j=0}^{\infty} \pi_j = 1 ,$$

which is called a <u>stationary</u> <u>distribution</u> for a Markov chain.

In the discussion below, we restrict ourselves to a Markov chain with finite state which are called a <u>finite</u> <u>Markov</u> <u>chain</u>. Applying Theorem (2.4.33) for a finite Markov chain, we can relabeling all the states and rewrite the transition probability matrix:

$$(2.4.37) \qquad \mathbf{P} = \begin{array}{c} C_1 \\ C_2 \\ \vdots \\ C_m \\ T \end{array} \left[\begin{array}{cccc|c} \mathbf{P}_1 & \cdots & \cdots & \mathbf{0} & \mathbf{0} \\ \cdots & \mathbf{P}_2 & \cdots & \mathbf{0} & \mathbf{0} \\ \cdots & \cdots & & \cdots & \cdots \\ \mathbf{0} & \cdots & & \mathbf{P}_m & \mathbf{0} \\ \hline \mathbf{R}_1 & \mathbf{R}_2 & \cdots & \mathbf{R}_m & \mathbf{Q} \end{array} \right]$$

where C_1, C_2, ..., C_m are all the sets of positive recurrent classes and T are a set of the remaining transient states. The limiting behavior of $p_{ij}^{\infty} = \lim_{n\to\infty} p_{ij}^{n}$ for all the states is summarized as follows:

$$(2.4.38) \qquad p_{ij}^{\infty} = 1/\mu_j \qquad (i, j \in C_k;\ k = 1, 2, \ldots, m),$$

$$(2.4.39) \qquad p_{ij}^{\infty} = f_{ij}/\mu_j \qquad (i \in T,\ j \in C_k;\ k = 1, 2, \ldots, m),$$

$$(2.4.40) \qquad p_{ij}^{\infty} = 0 \qquad (i \in C_k,\ j \in C_l;\ k \neq l),$$

$$(2.4.41) \qquad p_{ij}^{\infty} = 0 \qquad (i, j \in T),$$

$$(2.4.42) \qquad p_{ij}^{\infty} = 0 \qquad (i \in C_k;\ k = 1, 2, \ldots, m;\ j \in T),$$

where we assume that all the positive recurrent states are aperiodic. The above results are not perfect since we have to obtain f_{ij} for $i \varepsilon T$ and $j \varepsilon C_k$ $(k = 1, 2, \ldots, m)$.

Introducing a random variable $N_j(n)$ which is the number of visits to state j up to time n, we define

(2.4.43) $$E[N_j(\infty) | X(0)=i] = M_{ij} \quad (i, j \varepsilon T),$$

which is the expected number of visits to state j before moving to any recurrent state starting from state i at time 0.

(2.4.45) Theorem For a finite Markov chain with transition probability matrix (2.4.37), we have

(2.4.46) $$[M_{ij}] = [\mathbf{I} - \mathbf{Q}]^{-1} = \sum_{n=0}^{\infty} \mathbf{Q}^n,$$

where $\mathbf{I}$ is an identity matrix.

Note that $\mathbf{Q}$ is a square matrix describing the transitions among all the transient states for T, and there exists the inverse matrix $[\mathbf{I} - \mathbf{Q}]^{-1}$ since $\sum_{j\varepsilon T} p_{ij} < 1$ for at least any $i \varepsilon T$ and $\mathbf{Q}^n \to \mathbf{0}$ as $n \to \infty$.

(2.4.47) Theorem The first passage probability f_{ij} from a transient state i to a recurrent state j is given by

(2.4.48) $$[f_{ij}] = \sum_{l\varepsilon C_k} p_{il} + \sum_{l\varepsilon T} p_{il} f_{lj}$$

$$(i \varepsilon T, j \varepsilon C_k; k = 1, 2, \ldots, m)$$

and in a matrix form

$$(2.4.49) \qquad [f_{ij}] = [\mathbf{I} - \mathbf{Q}]^{-1}\mathbf{R}_k\mathbf{1} \qquad (i \in T,\ j \in C_k;\ k = 1, 2, \ldots, m),$$

where $\mathbf{1}$ is a column vector of all components unity.

Applying Theorem (2.4.47), we can calculate f_{ij} ($i \in T$, $j \in C_k$; $k = 1, 2, \ldots, m$) which is used for obtaining the limiting probability p_{ij}^n as $n \to \infty$ in (2.4.39).

We summarize the limiting transition probabilities for a finite Markov chain where all the recurrent states are assumed aperiodic: Relabeling all the states, we can obtain the transition probability matrix in (2.4.37). Calculating the stationary distributions $\{\pi_j, j \in C_k\}$ ($k = 1, 2, \ldots, m$) for all the recurrent classes and the first passage probability f_{ij} for $i \in T$, $j \in C_k$ ($k = 1, 2, \ldots, m$) in (2.4.49), we can finally calculate the limiting transition probabilities in (2.4.38) - (2.4.42).

2.5 Markov Processes

In the preceding section we have discussed a Markov chain which can move from one state to another at discrete-time. We can easily consider a continuous-time stochastic process which can move one state to another.

(2.5.1) Definition Consider a continuous-time stochastic process $\{X(t), t \geq 0\}$ with state space $i = 0, 1, 2, \ldots$.

For any $0 < t_1 < t_2 < \ldots < t_n < t$, if

$$(2.5.2) \qquad P\{X(t)=x|X(t_1)=x_1, X(t_2)=x_2, \ldots, X(t_n)=x_n\}$$
$$= P\{X(t)=x|X(t_n)=x_n\},$$

then the process $\{X(t), t \geq 0\}$ is called a Markov process.

That is, a Markov process is a stochastic process in which the probability law at time t is depending only on the present state $X(t_n)= x_n$, and is independent of the past history $X(t_1) = x_1, X(t_2) = x_2, \ldots, X(t_{n-1}) = x_{n-1}$.

Defining

$$(2.5.3) \qquad P_{ij}(t) = P\{X(t+s)=j|X(s)=i\} \quad (i, j = 0,1,2,\ldots)$$

for $t \geq 0$, $s \geq 0$, we are interested in the stationary transition probability $P_{ij}(t)$ which is independent of $s \geq 0$. Just similar to (2.4.6), we have

$$(2.5.4) \qquad P_{ij}(t) = \sum_{k=0}^{\infty} P_{ik}(t)P_{kj}(t) \quad (i, j = 0, 1, 2, \ldots)$$

which is called the <u>Chapman-Kolmogorov</u> <u>equation</u>. In a matrix form, we have

$$(2.5.5) \qquad \mathbf{P}(t+s) = \mathbf{P}(t)\mathbf{P}(s)$$

where $\mathbf{P}(t) = [P_{ij}(t)]$.

Before discussing the Markov processes in general, we first discuss the pure birth processes, and then the birth and death processes. We first discuss a pure birth process which is a counting process describing the number of events or births.

(2.5.6) Definition If a counting process $\{N(t),\ t \geq 0\}$ is a Markov process and the following conditions;

(i) $N(0) = 0$,
(ii) $P\{N(t+h) - N(t) = 1 | N(t) = k\} = \lambda_k h + o(h)$,
(iii) $P\{N(t+h) - N(t) \geq 2 | N(t) = k\} = o(h)$,

are satisfied, then the counting process $\{N(t),\ t \geq 0\}$ is called a <u>pure birth process</u> with parameters $\{\lambda_k,\ k = 0, 1, 2, \ldots\}$.

The pure birth process is a generalization of a Poisson process by allowing the parameter λ_k depending on the present state k $(k = 0, 1, 2, \ldots)$. Defining the probability

$$P_k(t) = P\{N(t) = k | N(0) = 0\} \quad (k = 0, 1, 2, \ldots), \tag{2.5.7}$$

we have the following differential equations:

$$P_0'(t) = -\lambda_0 P_0(t), \tag{2.5.8}$$

$$P_k'(t) = -\lambda_k P_k(t) + \lambda_{k-1} P_{k-1}(t) \quad (k = 1, 2, \ldots), \tag{2.5.9}$$

which can be similarly derived by specifying the times t and $t+h$ and tending to $h \to 0$. The differential equations (2.5.8) - (2.5.9) are called the <u>Kolmogorov's forward equations</u>, and can be solved recursively by

$$P_0(t) = e^{-\lambda_0 t}, \tag{2.5.10}$$

$$P_k(t) = \lambda_{k-1} e^{-\lambda_k t} \int_0^t e^{\lambda_k x} P_{k-1}(x)dx \quad (k = 1, 2, \ldots), \tag{2.5.11}$$

where the initial conditions $P_0(0) = 1$, and $P_k(0) = 0$ ($k = 1, 2, \ldots$) are applied.

The following theorem is clear if we refer to Theorem (2.2.18) for a Poisson process.

(2.5.12) Theorem For a pure birth process with parameters $\{\lambda_k, k = 0, 1, 2, \ldots\}$, the interarrival times X_k ($k = 1, 2, \ldots$) are independent and distributed exponentially with parameter λ_{k-1}, respectively.

Noting that the mean interarrival times are $E[X_k] = 1/\lambda_{k-1}$ ($k = 1, 2, \ldots$), and the expected time to infinitely many events are $\sum_{k=0}^{\infty} 1/\lambda_k$ which should be divergent, we have the following:

(2.5.13) Theorem For a pure birth process with parameters $\{\lambda_k, k = 0, 1, 2, \ldots\}$,

$$(2.5.14) \qquad \sum_{k=0}^{\infty} P_k(t) = 1$$

for all $t \geq 0$ holds if and only if

$$(2.5.15) \qquad \sum_{k=0}^{\infty} 1/\lambda_k = \infty.$$

It is quite true that $\sum_{k=0}^{\infty} 1/\lambda_k = \infty$ for a Poisson process since $\lambda_k = \lambda$ ($k = 0, 1, 2, \ldots$). We cite another example of the pure birth processes. A Yule process can be applied in biology, where the parameters $\lambda_k = (k+i)\lambda$ are prespecified (i; a positive integer). It is also true that $\sum_{k=0}^{\infty} 1/\lambda_k = \infty$ for the Yule process. We finally cite a counterexample of Theorem (2.5.13). If we specify the parameters $\lambda_k = 2^k \lambda$ ($\lambda > 0$, $k = 0, 1, 2, \ldots$) for the pure birth process, then $\sum_{k=0}^{\infty} 1/\lambda_k = 2/\lambda < \infty$ which implies that infinitely many events or births takes place for a finite time.

The pure birth process is a counting process describing the number of events or births of individuals (or events), then we can introduce the pure death process which can be similarly developed, except that state 0 is an absorbing state since no transitions can take place in state 0.

Combining the pure birth and pure death processes, we can introduce a birth and death process in which the sample function is shown in Fig. 2.5.1.

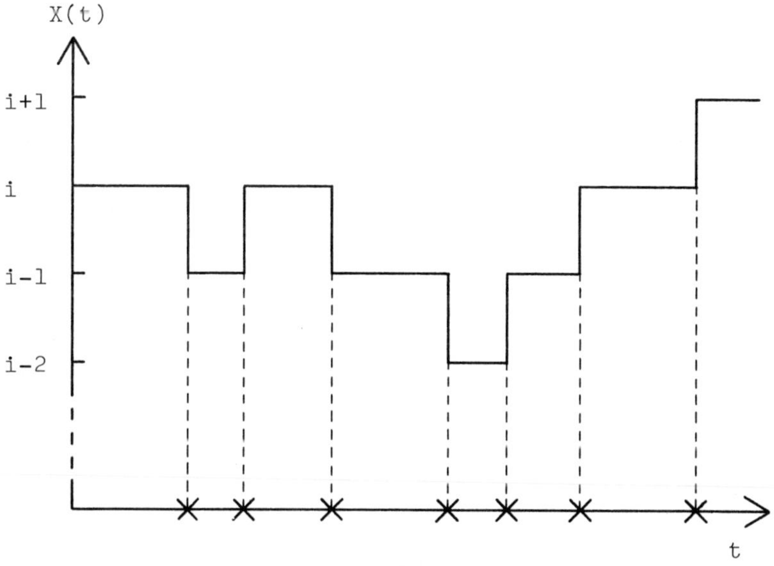

Fig. 2.5.1. A sample function of a birth and death process.

(2.5.16) Definition If a stochastic process $\{X(t), t \geq 0\}$ is a Markov process and the following conditions

(i) $X(0) = i$,
(ii) $P\{X(t+h) - X(t) = 1 | X(t) = k\} = \lambda_k h + o(h)$,
(iii) $P\{X(t+h) - X(t) = -1 | X(t) = k\} = \mu_k h + o(h)$,
(iv) $P\{2$ events take place in $[t, t+h] | X(t) = k\} = o(h)$,

are satisfied, then the process is called a <u>birth</u> <u>and</u> <u>death</u> <u>process</u> with parameters $\{\lambda_k, \mu_{k+1}, k = 0, 1, 2, \ldots\}$, where λ_k and μ_{k+1} are called the birth and death rates, respectively.

Introducing the transition probability $P_{ij}(t) = P\{X(t) = j | X(0) = i\}$ $(i, j = 0, 1, 2, \ldots)$, specifying the times t and $t+h$, and tending to $h \to 0$, we have

$$P'_{i0}(t) = -\lambda_0 P_{i0}(t) + \mu_1 P_{i1}(t) \quad (i = 0, 1, 2, \ldots) \tag{2.5.17}$$

$$P'_{ij}(t) = \lambda_{j-1} P_{i,j-1}(t) - (\lambda_j + \mu_j) P_{ij}(t) + \mu_{j+1} P_{i,j+1}(t) \quad (i=0,1,2,\ldots;j=1,2,\ldots), \tag{2.5.18}$$

which are called the <u>Kolmogorov's</u> <u>forward</u> <u>equations</u>, where the initial conditions are $P_{ii}(0) = 1$ and $P_{ij}(0) = 0$ $(i \neq j)$.

It is difficult or impossible except the simplest cases that the Kolmogorov's forward equations (2.5.17) - (2.5.18) can be solved analytically. However, we will later discuss the numerical computation of transition probabilities for the general Markov process.

A direct extension to Theorem (2.5.12) is the

following:

(2.5.19) Theorem For a birth and death process $\{X(t), t \geq 0\}$ with parameters $\{\lambda_k, \mu_{k+1}, k = 0, 1, 2, \ldots\}$, when the process is in state i $(i = 0, 1, 2, \ldots)$ at time t, the interarrival time to the next transition is independent of the other ones and distributed exponentially with parameter $\lambda_i + \mu_i$, where the probability of moving to the next state $i - 1$ $(i + 1)$ is $\mu_i/(\lambda_i + \mu_i)$ $(\lambda_i/(\lambda_i + \mu_i))$, respectively. Note that when $i = 0$, the possible transition state is only state 1 (that is, we interpret that $\mu_0 = 0$ and $\lambda_0 + \mu_0 = \lambda_0$ for state 0).

It is easier to discuss the asymptotic results as $t \to \infty$ for a birth and death process. We can show that $P'_{ij}(t)$ must be zero if there exist the limiting probabilities

$$(2.5.20) \qquad p_j = \lim_{t\to\infty} P_{ij}(t) \quad (i, j = 0, 1, 2, \ldots),$$

which are independent of the initial state i, and we tend to $t \to \infty$. Then from (2.5.17) - (2.5.18), we have the following simultaneous equations:

$$(2.5.21) \qquad -\lambda_0 p_0 + \mu_1 p_1 = 0,$$

$$(2.5.22) \qquad \lambda_{j-1}p_{j-1} - (\lambda_j + \mu_j)p_j + \mu_{j+1}p_{j+1} = 0 \qquad (j = 1, 2, \ldots),$$

$$(2.5.23) \qquad \sum_{j=0}^{\infty} p_j = 1 ,$$

which is the total probability. Solving the above simultaneous equations under the suitable conditions, we have the following:

(2.5.25) Theorem For a birth and death process with parameters $\{\lambda_k, \mu_{k+1}, k = 0, 1, 2, \ldots\}$, we assume that

$$\lambda_k > 0, \quad \mu_{k+1} > 0 \qquad (k = 0, 1, 2, \ldots), \tag{2.5.26}$$

there exist the limiting probabilities

$$p_j = \lim_{t \to \infty} P_{ij}(t) \quad (i, j = 0, 1, 2, \ldots) \tag{2.5.27}$$

which are independent of the initial state i if and only if

$$\sum_{j=1}^{\infty} \prod_{k=1}^{j} \frac{\lambda_{k-1}}{\mu_k} < \infty . \tag{2.5.28}$$

Then the limitng probabilities are given by

$$p_0 = \left(1 + \sum_{j=1}^{\infty} \prod_{k=1}^{j} \frac{\lambda_{k-1}}{\mu_k}\right)^{-1}, \tag{2.5.29}$$

$$p_j = \left(\prod_{k=1}^{j} \frac{\lambda_{k-1}}{\mu_k}\right) p_0 . \tag{2.5.30}$$

(2.5.31) Example As an example of the birth and death process, we cite an $M/M/1/\infty$ queue, where customers arrive at Poisson rate λ and are served exponentially at rate μ by a single channel, and the queue size may go to infinity. Then $\lambda_k = \lambda$, $\mu_{k+1} = \mu$ $(k = 0, 1, 2, \ldots)$ for a birth and death process. Applying the condition, we have

$$\sum_{j=1}^{\infty} \prod_{k=1}^{j} \frac{\lambda_{k-1}}{\mu_k} = \frac{\rho}{1 - \rho} < \infty \tag{2.5.32}$$

if and only if $\rho = \lambda/\mu < 1$. That is, if the arrival rate λ is less than the service rate μ, there exist the limiting probabilities

$$p_j = \rho^j (1 - \rho) \qquad (j = 0, 1, 2, \ldots). \tag{2.5.33}$$

Referring to Markov chain theory, we can classify all the states of a birth and death process as positive recurrent, null recurrent or transient. If all the parameters λ_k, μ_{k+1} $(k = 0, 1, 2, \ldots)$ are positive, all the states communicates each other, which implies that the birth and death process is irreducible.

If we further assume that $\lambda < \mu$ for the M/M/1/∞ queue, then the process is positive recurrent. If $\lambda = \mu$, then it is null recurrent. And if $\lambda > \mu$, then it is transient. There do not exist the limiting probabilities for the latter two cases (i.e., $\lambda = \mu$ and $\lambda > \mu$).

We are also interested in a birth and death process with finite state space. Let the state space be $i = 0, 1, 2, \ldots, m$, where m is finite. Then the Kolmogorov's forward equations are given by

(2.5.34) $$P'_{i0}(t) = -\lambda_0 P_{i0}(t) + \mu_1 P_{i1}(t),$$

(2.5.35) $$P'_{ij}(t) = \lambda_{j-1} P_{i,j-1}(t) - (\lambda_j + \mu_j) P_{ij}(t) + \mu_{j+1} P_{i,j+1}(t) \quad (j = 1, 2, \ldots, m-1),$$

(2.5.36) $$P'_{im}(t) = \lambda_{m-1} P_{i,m-1}(t) - \mu_m P_{im}(t).$$

We are ready to show the following:

(2.5.37) Theorem For a finite state birth and death process with parameters $\{\lambda_k, \mu_{k+1}, k = 0, 1, 2, \ldots, m-1\}$, we assume that

(2.5.38) $$\lambda_k > 0, \quad \mu_{k+1} > 0 \qquad (k = 0, 1, 2, \ldots, m-1).$$

Then there exist the limiting probabilities

$$(2.5.39) \qquad p_j = \lim_{t\to\infty} P_{ij}(t) = \begin{cases} \left(1 + \sum_{h=1}^{N} \prod_{k=1}^{h} \frac{\lambda_{k-1}}{\mu_k}\right)^{-1} & (j = 0) \\ \left(\prod_{k=1}^{j} \frac{\lambda_{k-1}}{\mu_k}\right) p_0 & (j = 1, 2, \ldots, m) \end{cases}$$

which are independent of the initial state i.

(2.5.40) Theorem Consider an M/M/1/N queue, where the arrival rate is λ , the service rate is μ , the service channel is a single, and the maximum system size for queue is $N < \infty$. Applying Theorem(2.5.37), we have the limiting probabilities

$$(2.5.41) \qquad p_j = (\lambda/\mu)^j [1 - (\lambda/\mu)]/[1 - (\lambda/\mu)^{N+1}] ,$$

where note that the condition such as equation (2.5.28) is not needed for a finite state birth and death process.

We are now ready to discuss the general Markov processes which can allow to move any state. In reliability applications it is enough to discuss the Markov processes with finite state space (i.e., finite Markov processes). We, therefore, restrict ourselves to discuss only the finite Markov processes with state space $i = 0, 1, \ldots, m$.

We assume that for a small time interval h,

$$(2.5.42) \qquad P_{ij}(h) = a_{ij}h + o(h) \quad (i \neq j)$$

where $a_{ij} > 0$ denotes the transition rate from state i to state j if it is possible, and $a_{ij} = 0$ if it is impossible to do so. Equation (2.5.41) can be rewritten by

$$\lim_{h\to 0} P_{ij}(h)/h = a_{ij} \qquad (i \neq j). \tag{2.5.43}$$

On the other hand, for a small time interval h,

$$1 - P_{ii}(h) = \sum_{\substack{j=0\\ j\neq i}}^{m} P_{ij}(h) = \sum_{\substack{j=0\\ j\neq i}}^{m} a_{ij}h + o(h), \tag{2.5.44}$$

or

$$\lim_{h\to 0} [1 - P_{ii}(h)]/h = \sum_{\substack{j=0\\ j\neq i}}^{m} a_{ij} = a_i, \tag{2.5.45}$$

where we define

$$a_i = \sum_{\substack{j=0\\ j\neq i}}^{m} a_{ij}. \tag{2.5.46}$$

Introducing the square matrix $\mathbf{A} = [a_{ij}]$ with non-diagonal element a_{ij} $(i \neq j)$ and diagonal element $-a_i$, we have from (2.5.42) and (2.5.43),

$$\lim_{h\to 0} [\mathbf{I} - \mathbf{P}(h)]/h = -\mathbf{A}, \tag{2.5.47}$$

or

$$\mathbf{P}(h) = \mathbf{I} + \mathbf{A}h + \mathbf{B}(o(h)), \tag{2.5.48}$$

where $\mathbf{B}(o(h))$ denotes a square matrix of all elements of $o(h)$. The matrix $\mathbf{A}$ is called the infinitestimal generator for the Markov process and each non-diagonal element denotes the transition rate. Applying the Chapman-Kolmogorov equation $\mathbf{P}(t+h) = \mathbf{P}(t)\mathbf{P}(h)$, we have

$$[\mathbf{P}(t+h) - \mathbf{P}(t)]/h = \mathbf{P}(t)\mathbf{A} + \mathbf{P}(t)\mathbf{B}(o(h))/h, \tag{2.5.49}$$

and tending to $h \to 0$, we have

$$(2.5.50) \qquad \mathbf{P}'(t) = \mathbf{P}(t)\mathbf{A},$$

which is the <u>Kolmogorov's</u> <u>forward</u> <u>equation</u> of matrix form for the Markov process.

In a similar fashion, applying the Chapman-Kolmogorov equation $\mathbf{P}(t+h) = \mathbf{P}(h)\mathbf{P}(t)$ and tending to $h \to 0$, we have

$$(2.5.51) \qquad \mathbf{P}'(t) = \mathbf{A}\mathbf{P}(t),$$

which is the <u>Kolmogorov's</u> <u>backward</u> <u>equation</u> of matrix form for the Markov process.

Using the initial condition $\mathbf{P}(0) = \mathbf{I}$, an identity matrix, and noting that the state space is finite, we have

$$(2.5.52) \qquad \mathbf{P}(t) = e^{\mathbf{A}t} = \mathbf{I} + \sum_{n=1}^{\infty} \mathbf{A}^n t^n / n!,$$

which is a unique solution both to the Kolmogorov's forward and backward equations in (2.5.50) and (2.5.51), respectively.

In principle, if we can derive all the eigen values and their corresponding row- and column-eigen vectors for matrix $\mathbf{A}$, we can calculate $\mathbf{P}(t)$ analytically. However, it is difficult or impossible to do so except the simplest cases.

The following method is one of the computational methods of $e^{\mathbf{A}t}$, which is called the <u>randomization</u> or <u>uniformization</u>. Let $\mathbf{Q}$ be the transformed matrix

(2.5.53) $\mathbf{Q} = \mathbf{A}/\Lambda + \mathbf{I}$

from $\mathbf{A}$, where

(2.5.54) $\Lambda = \max_i a_i$

Note that $\mathbf{Q}$ is the transition probability matrix for the discrete-time Markov chain since each row sum of $\mathbf{Q}$ is a unity and all the elements are non-negative. Note also that the structure of $\mathbf{A}$ is preserved for the transformed Markov chain $\mathbf{Q}$ (i.e., each state classification for $\mathbf{A}$ is preserved for $\mathbf{Q}$). Substituting $\mathbf{A} = \Lambda\mathbf{Q} - \Lambda\mathbf{I}$ (from (2.5.53)) into $\mathbf{P}(t) = e^{\mathbf{A}t}$, we have

(2.5.55) $$\mathbf{P}(t) = e^{\Lambda t\mathbf{Q}} e^{-\Lambda t\mathbf{I}}$$

$$= e^{-\Lambda t} e^{\Lambda t\mathbf{Q}}$$

$$= [\sum_{n=0}^{\infty} q_{ij}^{n} \; e^{-\Lambda t} \frac{(\Lambda t)^n}{n!}] \; ,$$

where $\mathbf{Q}^n = [q_{ij}^n]$, n-step transition probability matrix. The right-hand side of equation (2.5.55) is an infinite series of n-step transition probability by the Poisson probability mass function $e^{-\Lambda t}(\Lambda t)^n/n!$. Noting that $\mathbf{Q}^n$ converges the stationary probabilities for recurrent states and the unimodel property of the Poisson probability mass functions, we can compute (2.5.55) for a finite terms within a prespecified allowed error, say ε, instead of the infinite series.

As shown in Theorems (2.5.12) and (2.5.19), the interarrival times are independent and distributed exponentially. We can also show the following theorem for the general Markov process.

(2.5.56) Theorem Consider a finite Markov process whose transitions are described by the infinitestimal generator **A**. When the process is in state i (i = 0, 1, 2, ..., m), the interarrival time to the next transition is independent of other ones and distributed exponentially that the next state j to be visited is a_{ij}/a_i.

It is similar to classify each state of the Markov process by noting that $i \to j$ is equivalent to $a_{ij} > 0$. In the theorem below, we assume that all the states communicates each other, i.e., the Markov process is irreducible and positive recurrent (since the state space is finite). Note that the Markov process is aperiodic.

(2.5.57) Theorem If all the states communicate each other for a finite Markov process, there exists the limiting probabilities

(2.5.58) $$\lim_{t\to\infty} P_{ij}(t) = p_j > 0 \qquad (j = 0, 1, 2, \ldots, m)$$

which is independent of the initial state i. Let

(2.5.59) $$\mathbf{p} = [p_0, p_1, \ldots, p_m]$$

be the stationary distribution vector of the Markov process. Then **P** is a unique and positive solution to

(2.5.60) $$\mathbf{pA} = \mathbf{0},$$

(2.5.61) $$\sum_{j=0}^{m} p_j = 1.$$

Bibliography and Comments

Throughout all the sections: Many textbooks on stochastic processes have been published. See Karlin and Taylor (1975, 1981), Cinlar (1975) and Ross (a970, 1983). Stochastic models in reliability theory have been discussed by several authors. See Barlow and Proschan (1965, 1975), Mann, Shafer and Singpurwalla (1974), and Gnedenko, Belyayev and Solvyev (1965).

Section 2.5: The randomization or uniformization techniques for computing the transition probabilities for a Markov process are of great use for implementing computer programs: See Grassman (1977a, 1977b).

[1] R.E. Barlow and F. Proschan (1965), Mathematical Theory of Reliability, Wiley, New York.

[2] R.E. Barlow and F. Proschan (1975), Statistical Theory of Reliability and Life Testing - Probability Models, Holt, Rinehart and Winston, New York.

[3] E. Cinlar (1975), Introduction to Stochastic Processes, Prentice-Hall, Englewood Cliffs, N.J.

[4] B.V. Gnedenko, Yu. K. Belyayev, and A.D. Solvyev (1965), Mathematical Methods of Reliability Theory, Academic Press, New York.

[5] W.K. Grassman (1977a), "Transient Solutions in Markovian Queueing Systems," Computers and Operations Res., Vol. 4, pp. 47-53.

[6] W.K. Grassman (1977b), "Transient Solutions in Markovian Queues," European J. Operations Res., Vol. 1, pp. 396-402.

[7] S. Karlin and H.M. Taylor (1975), A First Course in Stochastic Processes, Academic Press, New York.

[8] S. Karlin and H.M. Taylor (1981), A Second Course in Stochastic Processes, Academic Press, New York.

[9] N.R. Mann, R.E. Shafer, and N.D. Singpurwalla (1974), Methods for Statistical Analysis of Reliability and Life Data, Wiley, New York.

[10] S.M. Ross (1970), Applied Probability Models with Optimization Applications, Holden-Day, San Francisco.

[11] S.M. Ross (1983), Stochastic Processes, Wiley, New York.

CHAPTER 3

MARKOV RENEWAL PROCESSES

3.1 Introduction

A Markov chain can move from one state to another at each fixed time interval. A Markov process can move from one state to another in which the sojourn time is distributed exponentially. Both a Markov chain and a Markov process can move from one state to another, in which the state space is finite or denumerable. On the other hand, a renewal process can revisit one state, in which the state space is only one. However, it can permit an arbitrary distribution of the sojourn time.

Combining a Markov chain and a renewal process implies a Markov renewal process or a semi-Markov process. A Markov renewal process is concerned with the generalized renewal random variables and a semi-Markov process is concerned with the random variables that the process is in a state at some time. However, both a Markov renewal process and a semi-Markov process are equivalent from the viewpoint of probability theory. In the sequel we shall use mainly a Markov renewal process.

In this chapter we summarize Markov renewal processes which can be applied to reliability modeling in the subsequent chapters. Section 3.2 is devoted to Markov renewal functions which are basic quantities in Markov renewal theory. Section 3.3 is devoted to stationary probabilities and asymptotic results. Section 3.4 is devoted to examples of Markov renewal processes which can be described by state transition diagrams (i.e., signal-flow graphs) and be solved graphically. Section 3.5 is devoted to Markov renewal processes with non-regeneration points which can be widely applied to reliability modeling in the subsequent chapters.

Consider a stochastic process $(\mathbf{X}, \mathbf{T})$ where the sample function is shown in Fig. 3.1.1. Let the state space be $i = 0, 1, 2, \ldots, m$, where m is finite unless otherwise mentioned. The vector random variable $\mathbf{X}$ is $\{X(n); n = 0, 1, 2, \ldots\}$, where $X(n) = i$ denotes that the process is in state i at discrete time n. The vector random variable $\mathbf{T}$ is $\{T_n; n = 0, 1, 2, \ldots\}$, where T_n denotes the random variable of the nth arrival time at which the process just moves from one state to another (possibly the same state). Specifying the random variables $\{Z(t), t \geq 0\}$, where $Z(t) = i$ denotes that the process is in state i at time t, and the family of the random variables $\{\mathbf{N}(t), t \geq 0\}$, where

$$(3.1.1) \qquad \mathbf{N}(t) = [N_0(t) \quad N_1(t) \quad \ldots \quad N_m(t)],$$

and $N_i(t) = k$ denotes that the random number of visits to state i is k in (0, t]. The random variable Z(t) can specify the history of all the states at any time, and the random variable **N**(t) can specify the generalized renewal random variable for all the states. We are ready to define a Markov renewal process and a semi-Markov process, respectively.

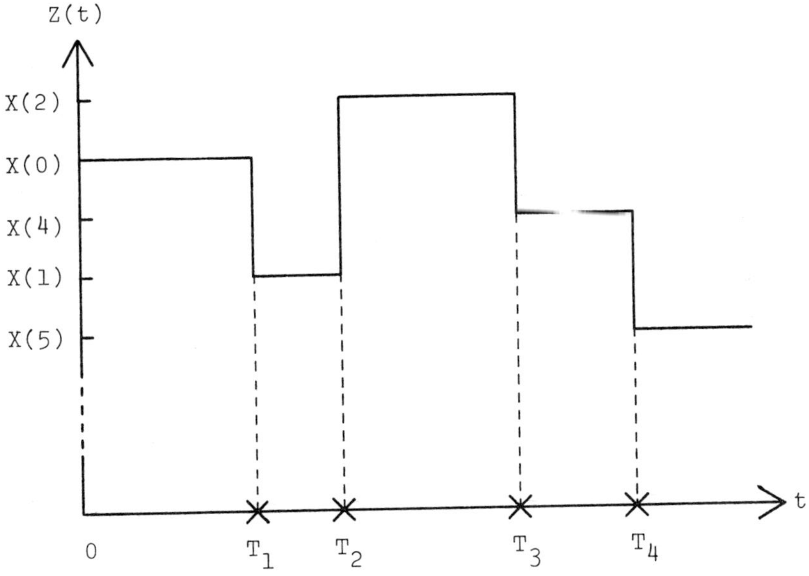

Fig. 3.1.1. A sample function of a Stochastic process (**X**, **T**).

(3.1.2) Definition A stochastic process $\{\mathbf{N}(t), t \geq 0\}$ is called a <u>Markov renewal process</u> if

$$(3.1.3) \quad P\{X(n+1)=j,\ T_{n+1}-T_n \le t \mid X(0)=i_0,\ \ldots,\ X(n)=i;\ T_0=0,\ T_1=t_1,\ \ldots,\ T_n=t_n\}$$

$$= P\{X(n+1)=j,\ T_{n+1}-T_n \le t \mid X(n)=i\}$$

$$= Q_{ij}(t)$$

is satisfied for all $n = 0, 1, 2, \ldots$; $i, j, = 0, 1, 2, \ldots, m$, and $t \in [0, \infty)$.

(3.1.4) Definition A stochastic process $\{Z(t), t \geq 0\}$ is called a semi-Markov process if equation (3.1.3) is satisfied.

Of course, we assume that the process is time-homogeneous, i.e.,

$$(3.1.5) \qquad Q_{ij}(t) = P\{X(n+1) = j,\ T_{n+1} - T_n \leq t \mid X(n) = i\}$$

is independent of n, where $Q_{ij}(t)$ is called a mass function or a one-step transition probability for the Markov renewal process and the matrix composed of $Q_{ij}(t)$,

$$(3.1.6) \qquad \mathbf{Q}(t) = [Q_{ij}(t)]$$

is called a semi-Markov kernel. The reason that $Q_{ij}(t)$ is called a one-step transition probability is the following: $Q_{ij}(t)$ is the probability that, after making a transition into state i, the process next makes a transition into state j, in an amount of time less than or equal to t. The one-step transition probabilities satisfy the following:

$$(3.1.7) \qquad Q_{ij}(t) \geq 0 \qquad (i, j = 0, 1, 2, \ldots, m),$$

(3.1.8) $$\sum_{j=0}^{m} Q_{ij}(\infty) = 1 \qquad (i = 0, 1, 2, \ldots, m).$$

Let

(3.1.9) $$p_{ij} = \lim_{t\to\infty} Q_{ij}(t) = P\{X(n+1) = j \mid X(n) = i\}$$

denote the eventual transition probability that the process can move from state i to state j. That is, p_{ij} is the eventual transition probability, neglecting the sojourn time, which governs the behavior of the random variable $\{X(n), n \geq 0\}$. The eventual transition probabilities satisfy the following:

(3.1.10) $$p_{ij} \geq 0 \qquad (i, j = 0, 1, 2, \ldots, m),$$

(3.1.11) $$\sum_{j=0}^{m} p_{ij} = 1 \quad (i = 0, 1, 2, \ldots, m).$$

A Markov chain $\{X(n), n = 0, 1, 2, \ldots\}$ having the eventual transition probabilities is called an <u>embedded Markov chain</u>. If we assume all the non-pathological distributions (i.e., all the non-zero sojourn time distributions), the behavior of the Markov renewal process is governed by the embedded Markov chain by ignoring all the sojourn times among the states.

If $p_{ij} > 0$ for some i and j, then we can define

(3.1.12) $$F_{ij}(t) = Q_{ij}(t)/p_{ij} ,$$

and if $p_{ij} = 0$ for some i and j, then we can define $Q_{ij}(t) = 0$ for all t and $F_{ij}(t) = 1(t)$ (a step function) in (3.1.12). The distribution $F_{ij}(t)$ is a distribution of the sojourn time that the process spends in state i given that the next visiting state is j.

(3.1.13) Example Consider a Markov renewal process $\{N_0(t), t \geq 0\}$ with one state $i = 0$, where $p_{00} = 1$ and $F_{00}(t)$ is an arbitrary distribution. That is, such a Markov renewal process is a renewal process with inter-arrival time distribution $F_{00}(t)$, which has been thoroughly discussed in Section 2.3.

(3.1.14) Example Consider a semi-Markov process $\{Z(t), t \geq 0\}$, where we assume that $Q_{ij}(t) = p_{ij}1(t-1)$ for $i, j = 0, 1, \ldots, m$, where $1(t-1)$ is a step function at $t = 1$. Then such a semi-Markov process $\{Z(t), t \geq 0\}$ is a discrete-time Markov chain $\{Z(n), n = 0, 1, 2, \ldots\}$ with transition probabilities p_{ij} $(i, j = 0, 1, \ldots, m)$.

(3.1.15) Example Consider a semi-Markov process $\{Z(t), t \geq 0\}$, where we assume that

$$(3.1.16) \qquad Q_{ij}(t) = p_{ij}(1 - e^{-\lambda_i t})$$

if $p_{ij} > 0$ $(i, j = 0, 1, 2, \ldots, m;\ i \neq j)$ and

$$(3.1.17) \qquad Q_{ij}(t) = 0$$

if $p_{ij} = 0$ $(i, j = 0, 1, 2, \ldots, m;\ i \neq j)$ and $Q_{ii}(t) = 0$ $(i = 0, 1, 2, \ldots, m)$. Then such a semi-Markov process is a continuous-time Markov process with the infinitestimal generator **A** with its elements $p_{ij}\lambda_i$ for non-diagonal and $-\lambda_i$ for diagonal.

As shown in the above examples, a Markov renewal process or a semi-Markov process includes a renewal process, a Markov chain and a Markov process as its special cases.

That is, a Markov renewal process is a generalization of a renewal process, a Markov chain and a Markov process, can permit the arbitrary sojourn time distributions $F_{ij}(t)$ for all i and j, and satisfies the Markov property in (3.1.3) in a sense. The Markov property in (3.1.3) shows us that the process is governed by the elapsed time from the latest time instant at which a transition (i.e., an event or occurrence) takes place. Such a time instant at which a transition takes place is called a <u>regeneration point</u>. Following to the successive regeneration points (i.e., T_n, n = 1, 2, ..., in Fig. 3.1.1), we can analyze a Markov renewal process by applying and expanding the existing renewal and Markov chain theories, since the behavior of the process can be determined by the latest regeneration point which can specify the elapsed time and the latest state visited.

Cinlar (1975) gave a classification of states for a Markov renewal process by considering the corresponding renewal process. However, we can simply give the following classification with an exception: The classification is just same to that of the corresponding embedded Markov chain. However, the periodicity has nothing to do with that of the embedded Markov chain. If the embedded Markov chain is irreducible, then all the states are aperiodic or they are all periodic. However, the latter is not periodic if two different $Q_{ij}(t)$ and $Q_{kl}(t)$ $(i \neq k)$ are lattice with respective different periods whose greatest common divisor is one. We will show an exception. If the embedded Markov chain is irreducible and every $Q_{ij}(t)$ (i, j = 0, 1, ..., m) is lattice with common period δ, then such a Markov renewal process is called a discrete-time Markov renewal process which has some applications in reliability theory.

Before concluding this section, we should note the following: If the sojourn time $F_{ij}(t)$ is distributed exponentially, then the process satisfies the Markov property (3.1.3). In this case, the process is independent of the elapsed time t in (3.1.3) because of the memoryless property of the exponential distribution. However, such a time instant is no longer a regeneration point. A regeneration point is a time instant at which a transition takes place even though it is independent of the elapsed time up to that time instant. Since we are concerned with the renewal random variable $\mathbf{N}(t)$ which can count the numbers of events or occurrences in the process for $t \varepsilon (0, t]$, we have to specify the regeneration points at which not only the Markov property in (3.1.3) is satisfied, but also an event or occurrence takes place (refer to a Poisson process in Section 2.2).

3.2 Markov Renewal Functions

Recall that $Q_{ij}(t) = p_{ij}F_{ij}(t)$ is a one-step transition probability from state i to state j for i, j = 0, 1, 2, ..., m. Assume that the first and second moments of the sojourn time distribution $F_{ij}(t)$ exist and are given by

(3.2.1) $$\nu_{ij} = \int_0^\infty t dF_{ij}(t),$$

(3.2.2) $$\nu_{ij}^{(2)} = \int_0^\infty t^2 dF_{ij}(t),$$

respectively. Define

(3.2.3) $$H_i(t) = \sum_{j=0}^{m} Q_{ij}(t) \qquad (i = 0, 1, 2, \ldots, m)$$

which is called the unconditional distribution in state i since $H_i(t)$ is the distribution not specifying the next visiting state. We also define the first and second moments of $H_i(t)$:

(3.2.4) $$\xi_i = \int_0^\infty t dH_i(t) = \sum_{j=0}^{m} p_{ij}\nu_{ij} \quad (i = 0, 1, 2, \ldots, m),$$

(3.2.5) $$\xi_i^{(2)} = \int_0^\infty t^2 dH_i(t) = \sum_{j=0}^{m} p_{ij}\nu_{ij}^{(2)} \quad (i = 0, 1, 2, \ldots m),$$

respectively. In particular, ξ_i is called the unconditional mean in state i.

Define the Markov renewal function

(3.2.6) $$M_{ij}(t) = E[N_j(t)|Z(0)=i] \quad (i,j = 0, 1, 2, \ldots, m),$$

which is the generalized renewal function and is the expected number of visits to state j in an interval (0, t] given that the process started in state i at time 0. Combining renewal and Markov chain theories imply

(3.2.7) $$\begin{aligned} M_{ij}(t) &= \sum_{\substack{k=0 \\ k\neq j}}^{\infty} \int_0^t M_{kj}(t-x)dQ_{ik}(x) + \int_0^t [1 + M_{jj}(t-x)]dQ_{ij}(x) \\ &= Q_{ij}(t) + \sum_{k=0}^{m} \int_0^t M_{kj}(t-x)dQ_{ik}(x) \\ &= Q_{ij}(t) + \sum_{k=0}^{m} Q_{ik}*M_{kj}(t), \end{aligned}$$

where the notation $*$ denotes the Stieltjes convolution. Introducing the matrices $\mathbf{M}(t) = [M_{ij}(t)]$ and $\mathbf{Q}(t) = [Q_{ij}(t)]$ in (3.1.6), we have the following matrix form:

$$\mathbf{M}(t) = \mathbf{Q}(t) + \mathbf{Q}*\mathbf{M}(t), \tag{3.2.8}$$

where the notation $*$ in matrix denotes the matrix multiplication except the multiplication for each element is replaced by the Stieltjes convolution for each element. Equation (3.2.8) is a renewal equation in a matrix form which is a direct extension in (2.3.21).

Equation (3.2.8) can be rewritten by

$$[\mathbf{I} - \mathbf{Q}]*\mathbf{M}(t) = \mathbf{Q}(t), \tag{3.2.9}$$

where $\mathbf{I}$ is an identity matrix with the diagonal elements $1(t)$ (a step function). Noting that the inverse $[\mathbf{I} - \mathbf{Q}(t)]^{(-1)*}$ exists for a finite t and is given by

$$[\mathbf{I} - \mathbf{Q}(t)]^{(-1)*} = \sum_{n=0}^{\infty} \mathbf{Q}^{n*}(t), \tag{3.2.10}$$

we have

$$\begin{aligned} \mathbf{M}(t) &= [\mathbf{I} - \mathbf{Q}]^{(-1)*}*\mathbf{Q}(t) \\ &= \sum_{n=1}^{\infty} \mathbf{Q}^{n*}(t) \\ &= [\mathbf{I} - \mathbf{Q}(t)]^{(-1)*} - \mathbf{I} \end{aligned} \tag{3.2.11}$$

which corresponds to equation (2.3.21) in a matrix form.

Let $P_{ij}(t)$ denote the transition probability that the

process is in state j at time t given that it was in state i at time 0:

$$P_{ij}(t) = P\{Z(t) = j \mid Z(0) = i\}. \quad (3.2.12)$$

Let $G_{ij}(t)$ denote the first passage time distribution that the process first arrives at state j at time t given that it was in state i at time 0:

$$G_{ij}(t) = P\{N_j(t) > 0 \mid Z(0) = i\}. \quad (3.2.13)$$

The transition probability $P_{ij}(t)$ can be written in terms of $Q_{ij}(t)$ and $H_i(t)$:

$$P_{ij}(t) = [1 - H_i(t)]\delta_{ij} + \sum_{k=0}^{m} Q_{ik} * P_{kj}(t). \quad (3.2.14)$$

That is, if $i \neq j$, the process can move to state k and then move to state j in an amount of time t. If $i = j$, there is another possibility that the process can also stay in state i in an amount of time t, which can be written by the first term of the right-hand side of equation (3.2.14), where δ_{ij} is a Kronecker's delta (i.e., $\delta_{ij} = 1$ if $i = j$ and $\delta_{ij} = 0$ if $i \neq j$).

The transition probability $P_{ij}(t)$ can be also written in terms of $Q_{ij}(t)$ and $G_{ij}(t)$:

$$P_{ij}(t) = [1 - H_i(t)]\delta_{ij} + G_{ij} * P_{jj}(t) \quad (i, j = 0, 1, 2, \ldots, m), \quad (3.2.15)$$

which can be similarly interpreted: That is, if $i \neq j$, the process can first move to state j and then follow the transition probability $P_{jj}(t)$. If $i = j$, there is another possibility that the process can also stay in state i in an amount of time t.

The first passage time distribution $G_{ij}(t)$ can be written in terms of $Q_{ij}(t)$:

$$(3.2.16) \qquad G_{ij}(t) = Q_{ij}(t) + \sum_{\substack{k=0 \\ k\neq j}}^{m} Q_{ik}*G_{kj}(t),$$

which can be similarly interpreted.

Let us introduce the Laplace-Stieltjes transform:

$$(3.2.17) \quad Q_{ij}^{*}(s) = \int_0^{\infty} e^{-st} dQ_{ij}(t) \quad (i,\ j = 0,\ 1,\ 2,\ \ldots,\ m)$$

and the following matrix of its element $Q_{ij}^{*}(s)$:

$$(3.2.18) \qquad \mathbf{Q}^{*}(s) = [Q_{ij}^{*}(s)].$$

We also introduce the Laplace-Stieltjes transform:

$$(3.2.19) \quad M_{ij}^{*}(s) = \int_0^{\infty} e^{-st} dM_{ij}(t) \quad (i,\ j = 0,\ 1,\ 2,\ \ldots,\ m)$$

and the matrix $\mathbf{M}^{*}(s) = [M_{ij}^{*}(s)]$.

Making the Laplace-Stieltjes transforms in (3.2.8), we have

$$(3.2.20) \qquad \mathbf{M}^{*}(s) = \mathbf{Q}^{*}(s) + \mathbf{Q}^{*}(s)\mathbf{M}^{*}(s).$$

Introducing an identity matrix $\mathbf{I}$ and noting that the inverse $[\mathbf{I} - \mathbf{Q}^{*}(s)]^{-1}$ exists for $\mathrm{Re}(s) > 0$, we have

$$(3.2.21) \qquad \mathbf{M}^*(s) = [\mathbf{I} - \mathbf{Q}^*(s)]^{-1}\mathbf{Q}^*(s)$$

$$= [\mathbf{I} - \mathbf{Q}^*(s)]^{-1} - \mathbf{I},$$

which corresponds to (2.3.25) in a matrix form. Once $\mathbf{Q}(t)$ is given, we can obtain the matrix Laplace-Stieltjes transform $\mathbf{Q}^*(s)$ and its inverse $[\mathbf{I} - \mathbf{Q}^*(s)]^{-1}$. That is, $\mathbf{M}^*(s)$ can be given by manipulating the inverse matrix. The Laplace-Stieltjes transforms

$$(3.2.22) \qquad G^*_{ij}(s) = \int_0^\infty e^{-st}dG_{ij}(t) \quad (i, j = 0, 1, 2, \ldots, m)$$

and

$$(3.2.23) \qquad P^*_{ij}(s) = \int_0^\infty e^{-st}dP_{ij}(t) \quad (i, j = 0, 1, 2, \ldots, m)$$

can be recursively given as follows: Noting that

$$(3.2.24) \qquad M^*_{ij}(s) = G^*_{ij}(s) + G^*_{ij}(s)M^*_{jj}(s),$$

we have

$$(3.2.25) \qquad G^*_{ij}(s) = M^*_{ij}(s)/[1 + M^*_{jj}(s)],$$

and

$$(3.2.26) \qquad P^*_{jj}(s) = [1 - H^*_j(s)]/[1 - G^*_{jj}(s)],$$

$$(3.2.27) \qquad P^*_{ij}(s) = G^*_{ij}(s)P^*_{jj}(s) \qquad (i \neq j).$$

We can analytically obtain the Laplace-Stieltjes transforms $M^*_{ij}(s)$ in (3.2.21), $G^*_{ij}(s)$ in (3.2.25), and $P^*_{ij}(s)$ in (3.2.26) and (3.2.27), respectively. In principle, we can invert the corresponding Laplace-Stieltjes transforms

$M_{ij}^*(s)$, and $G_{ij}^*(s)$ and $P_{ij}^*(s)$, which imply the analytical forms of $M_{ij}(t)$, $G_{ij}(t)$ and $P_{ij}(t)$. However, it is very difficult or impossible to do so except the simplest cases.

3.3 Stationary Probabilities

Barlow and Proschan (1965) showed that the first and second moments of $G_{ij}(t)$, denoted by μ_{ij} and $\mu_{ij}^{(2)}$, respectively, satisfy

$$(3.3.1) \qquad \mu_{ij} = \sum_{k \neq j} p_{ik}(\nu_{ik} + \mu_{kj}) + p_{ij}\nu_{ij}$$

$$= \sum_{k \neq j} p_{ik}\mu_{kj} + \xi_i,$$

and

$$(3.3.2) \qquad \mu_{ij}^{(2)} = \xi_i^{(2)} + \sum_{k \neq j} p_{ik}[\mu_{kj}^{(2)} + 2\nu_{ik}\,\mu_{kj}],$$

where ξ_i and $\xi_i^{(2)}$ are the first and second moments of $H_i(t)$, which are given in (3.2.4) and (3.2.5), and ν_{ij} is the mean of $F_{ij}(t)$, which is given in (3.2.1). If the embedded Markov chain is irreducible and aperiodic, there exists a unique and positive stationary probability vector $\underline{\pi} = [\pi_j]$, where $\underline{\pi}$ satisfies

$$(3.3.3) \qquad \underline{\pi}\mathbf{P} = \underline{\pi}$$

$$(3.3.4) \qquad \sum_{i=0}^{m} \pi_i = 1.$$

Note that $\mathbf{P} = [p_{ij}]$ is the transition probability matrix of the embedded Markov chain. Multiplying π_i to both sides and summing over i in (3.3.1) and (3.3.2), we have

$$\mu_{jj} = \sum_{k=0}^{m} \pi_k \xi_k / \pi_j, \tag{3.3.5}$$

$$\mu_{jj}^{(2)} = (1 + \sum_{k \neq j} \pi_k \mu_{kj}) / \pi_j. \tag{3.3.6}$$

The probability $G_{ij}(\infty)$ that the process is ultimately in state j given that it was in state i at time 0 is given by

$$G_{ij}(\infty) = \begin{cases} 1 & (i,\ j \ \varepsilon\ C_k;\ k = 1,\ 2,\ \ldots,\ K) \\ 0 & (i \ \varepsilon\ C_k,\ j \ \varepsilon\ C_l;\ k \neq l) \\ f_{ij} & (i \ \varepsilon\ T,\ j \ \varepsilon\ C_k;\ k = 1,\ 2,\ \ldots,\ K), \end{cases} \tag{3.3.7}$$

where f_{ij} is given in (2.4.48) or (2.4.49) for a discrete-time finite Markov chain whose transition probability matrix is given by $\mathbf{P} = [p_{ij}]$, that of the embedded Markov chain and whose states can be classified into some recurrent classes $C_1, C_2, \ldots, C_K$ and a set T of the remaining states which are transient (see (2.4.37)). These results are just the same of the discrete-time Markov chain theory developed in Section 2.4 since we are concerned with the transition probability and can neglect the elapsed time in each state.

We are now ready to derive the stationary probabilities (j = 0, 1, 2, ..., m) for a Markov renewal process. We first note that the stationary probabilities π_j (j = 0, 1, 2, ..., m) of the embedded Markov chain can be given by the discrete-time Markov chain theory developed in Section 2.4, and then we should derive the stationary probabilities ρ_j for the Markov renewal process.

If we assume that the Markov renewal process is irreducible and aperiodic, we apply the Tauberian theorem to (3.2.26):

$$(3.3.8)\quad \lim_{t\to\infty} P_{jj}(t) = \lim_{s\to 0} P^{*}_{jj}(s) = \lim_{s\to 0} \frac{[1 - H^{*}_{j}(s)]/s}{[1 - G^{*}_{jj}(s)]/s} = \frac{\xi_j}{\mu_{jj}}.$$

In general, we have from (3.2.27):

$$(3.3.9)\qquad \lim_{t\to\infty} P_{ij}(t) = \lim_{s\to 0} G^{*}_{ij}(s)P^{*}_{jj}(s) = G_{ij}(\infty)\ \xi_j/\mu_{jj}.$$

Summarizing the above results, we have the following:

(3.3.10) Theorem The stationary probabilities ρ_j $(j = 0, 1, 2, \ldots, m)$ for a Markov renewal process are given by

$$(3.3.11)\quad \rho_j = \begin{cases} \xi_j/\mu_{jj} & (j \in C_k;\ k = 1, 2, \ldots, K) \\ G_{ij}(\infty)\xi_j/\mu_{jj} & (i \in T,\ j \in C_k;\ k = 1, 2, \ldots, K) \\ 0 & (\text{otherwise}), \end{cases}$$

if $\xi_j < \infty$ for all $j = 0, 1, 2, \ldots, m$.

In particular, if the process is irreducible and aperiodic, we have

$$(3.3.12)\quad \rho_j = \xi_j/\mu_{jj} = \pi_j\xi_j/\sum_{k=0}^{m}\pi_k\xi_k \quad (j = 0, 1, 2, \ldots, m)$$

from (3.3.5).

Let $T_{N(t)}$ be the arrival time of the last transition

before or at time t and $T_{N(t)+1}$ be the next arrival time after t. Then the random variables

(3.3.13) $$\gamma(t) = T_{N(t)+1} - t$$

and

(3.3.14) $$\delta(t) = t - T_{N(t)}$$

are of interest, where $\gamma(t)$ and $\delta(t)$ are the generalized excess (residual life) and shortage (age) random variables, respectively (see Fig. 3.3.1). If we assume that the Markov renewal process is irreducible and aperiodic, and the mean recurrence times μ_{jj} are finite, we have the following limiting results:

(3.3.15) $$\lim_{t\to\infty} P\{\delta(t) \le x | Z(t) = i\}$$
$$= \lim_{t\to\infty} P\{\gamma(t) \le x | Z(t) = i\}$$
$$= (1/\xi_i)\int_0^x [1 - H_i(u)]du,$$

which generalizes the well-known result (2.3.54) for a renewal process.

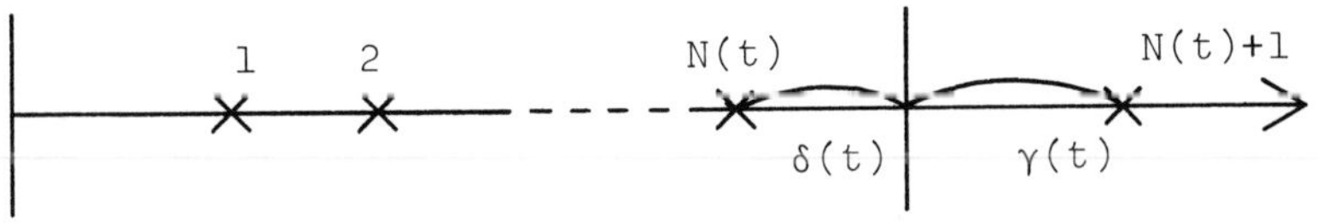

Fig. 3.3.1. A relationship among t, $N(t)$ and $N(t) + 1$.

Let us define the following probability:

(3.3.16) $\quad {}_iR_{jk}(x;t)$

$$= P\{Z(t)= j,\ X(N(t)+1)=k,\ T_{N(t)+1}-t\leq x\,|\,Z(0)=i\}$$
$$(i,\ j,\ k = 0,\ 1,\ 2,\ \ldots,\ m).$$

We are concerned with the limiting probability as $t \to \infty$. Pyke (1961b) showed the following:

(3.3.17) Theorem If state j is recurrent and aperiodic and $\nu_{jk} < \infty$, then

(3.3.18) $$\lim_{t\to\infty} {}_iR_{jk}(x;t) = G_{ij}(\infty)p_{jk}\mu_{jj}^{-1} \int_0^t [1 - F_{jk}(y)]dy.$$

Pyke (1961b) further showed that the stationary Markov renewal process can be constructed from the stationary probabilities ρ_j $(j = 0, 1, 2, \ldots, m)$ in (3.3.12) and $\lim_{t\to\infty} {}_iR_{jk}(x;t)$ in (3.3.18) if we assume that the process is irreducible and aperiodic. That is, if we assume that the process is irreducible and aperiodic, we can consider the following process:

(3.3.19) $$P\{Z(0) = j\} = \rho_j \qquad (j = 0,\ 1,\ 2,\ \ldots,\ m),$$

(3.3.20) $$P\{X(1) = j,\ T_1 - T_0 \leq x\,|\,X(0) = i\}$$
$$= p_{ij}\xi_i^{-1}\int_0^x [1 - F_{ij}(y)]dy,$$

(3.3.21) $$P\{X(n+1) = j,\ T_{n+1} - T_n \leq x\,|\,X(n) = i\} = Q_{ij}(t)$$
$$(n = 1,\ 2,\ \ldots).$$

We can construct the stationary Markov renewal process which is a direct expansion of a stationary renewal process discussed in Section 2.3.

Let us discuss the asymptotic behavior of $M_{ij}(t)$. From (3.2.24), we have

$$(3.3.22) \qquad M_{ij}^*(s) = G_{ij}^*(s)[1 + M_{jj}^*(s)] = G_{ij}^*(s)\left[1 + \frac{G_{jj}^*(s)}{1 - G_{jj}^*(s)}\right] .$$

Noting that

$$(3.3.23) \qquad G_{ij}^*(s) = 1 - s\mu_{ij} + \frac{s^2}{2!}\mu_{ij}^{(2)} + o(s^2) ,$$

and substituting a Taylor series of $G_{ij}^*(s)$ into (3.3.22), we have

$$(3.3.24) \qquad M_{ij}^*(s) = \frac{1}{s\mu_{jj}} + \frac{\mu_{jj}^{(2)}}{2\mu_{jj}^2} - \frac{\mu_{ij}}{\mu_{jj}} + o(1) ,$$

if we assume that states i and j belong to the same recuurent class. Applying Tauberian theorem, we have

$$(3.3.25) \qquad M_{ij}(t) - \frac{t}{\mu_{jj}} \to \frac{\mu_{jj}^{(2)}}{2\mu_{jj}^2} - \frac{\mu_{ij}}{\mu_{jj}} \quad \text{as } t \to \infty ,$$

which can be obtained by analogy with a delayed renewal process since the first interarrival time is distributed with $G_{ij}(t)$ and the successive interarrival times are distributed identically with $G_{jj}(t)$.

3.4 Examples and Signal-Flow Graphs

In this section we discuss some examples of Markov renewal processes and derive the probabilistic quantities obtained in the preceding sections.

The first example is an alternating renewal process whose state space is composed of two states, $i = 0, 1$, say, where state 0 is an operating state and state 1 is a failed state. The process can move from one state to another according to arbitrary distributions. Let $\{\mathbf{N}(t), t \geq 0\}$ be a Markov renewal process of an alternating renewal process, whose semi-Markov kernel is given by

$$(3.4.1) \qquad \mathbf{Q}(t) = \begin{matrix} 0 \\ 1 \end{matrix} \begin{bmatrix} 0 & F(t) \\ G(t) & 0 \end{bmatrix},$$

where $F(t)$ and $G(t)$ are arbitrary distributions of operating and failed states, respectively. It is clear that

$$(3.4.2) \qquad \mathbf{M}(t) = \begin{bmatrix} \sum_{n=1}^{\infty} (F*G)^{n*}(t) & \sum_{n=0}^{\infty} (F*G)^{n*}*F(t) \\ \sum_{n=0}^{\infty} (F*G)^{n*}*G(t) & \sum_{n=1}^{\infty} (F*G)^{n*}(t) \end{bmatrix},$$

and its Laplace-Stieltjes transform is given by

$$(3.4.3) \qquad \mathbf{M}^*(s) = \frac{1}{1 - F^*(s)G^*(s)} \begin{bmatrix} F^*(s)G^*(s) & F^*(s) \\ G^*(s) & F^*(s)G^*(s) \end{bmatrix}$$

The first passage time distributions are given by

(3.4.4) $$G_{00}(t) = G_{11}(t) = F*G(t),$$

(3.4.5) $$G_{01}(t) = F(t),$$

(3.4.6) $$G_{10}(t) = G(t).$$

The transition probabilities are given by

(3.4.7) $$P_{00}(t) = [1 - F]*\sum_{n=0}^{\infty} (F*G)^{n*}(t),$$

(3.4.8) $$P_{11}(t) = [1 - G]*\sum_{n=0}^{\infty} (F*G)^{n*}(t),$$

(3.4.9) $$P_{01}(t) = F*[1 - G]*\sum_{n=0}^{\infty} (F*G)^{n*}(t),$$

(3.4.10) $$P_{10}(t) = G*[1 - F]*\sum_{n=0}^{\infty} (F*G)^{n*}(t).$$

These probabilistic quantities can be directly derived since the process visits one state to another alternately. For instance, the sample functions for deriving $P_{00}(t)$ is shown in Fig. 3.4.1. That is, Fig. 3.4.1 (a) shows that the process stays in state 0 for an interval $(0, t]$, (b) shows that the process is in state 0 at time t after visiting state 1 once, and (c) shows that the process is in state 0 at time t after visiting state 1 n times, where $n = 2, 3, \ldots$. Then

(3.4.11) $$P_{00}(t) = [1 - F(t)] + F*G*[1 - F](t) + \sum_{n=2}^{\infty} (F*G)^{n*}*[1 - F](t).$$

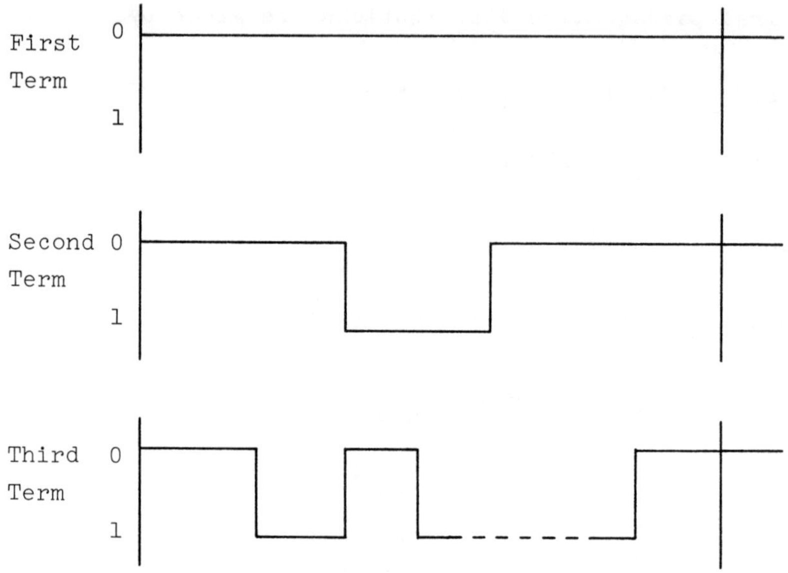

Fig. 3.4.1. Sample functions for deriving $P_{00}(t)$.

Let us consider the limiting behavior of the alternating renewal process. Assume that the process is aperiodic. If F(t) and G(t) are lattice with same period δ, then the process is periodic. Otherwise, the process is aperiodic. We restrict ourselves to the latter. The embedded Markov chain is given by

$$(3.4.12) \qquad \mathbf{P} = \begin{bmatrix} 0 & 1 \\ 1 & 0 \end{bmatrix}$$

and the limiting distribution of the embedded Markov chain is given by

$$(3.4.13) \qquad \pi_0 = \pi_1 = 1/2,$$

and the unconditional means for states 1 and 2 are defined by

(3.4.14) $\int_0^\infty t dF(t) = 1/\lambda,$

(3.4.15) $\int_0^\infty t dG(t) = 1/\mu,$

respectively. Then the limiting probabilities of the alternating renewal process are

(3.4.16) $$\rho_0 = \frac{(1/2)(1/\lambda)}{(1/2)(1/\lambda) + (1/2)(1/\mu)} = \frac{\mu}{\lambda + \mu},$$

(3.4.17) $$\rho_1 = \frac{(1/2)(1/\mu)}{(1/2)(1/\lambda) + (1/2)(1/\mu)} = \frac{\lambda}{\lambda + \mu}.$$

We finally derive the stationary Markov renewal process for the alternating renewal process:

(3.4.18) $P\{Z(0) = j\} = \rho_j \quad (j = 0, 1),$

(3.4.19) $P\{X(1) = 1, T_1 - T_0 \le x | X(0) = 0\}$

$= \lambda\int_0^x [1 - F(t)]dt = F_e(x),$

(3.4.20) $P\{X(1) = 0, T_1 - T_0 \le x | X(0) = 1\}$

$= \mu\int_0^x [1 - G(t)]dt = G_e(x),$

and

(3.4.21) $P\{X(n+1) = j, T_{n+1} - T_n \le x | X(n) = i\} = Q_{ij}(x)$
$(n = 1, 2, \ldots).$

Then the stationary process is constructed. For instance. if the process is in state 0 at time t given that it started from states 0 and 1, we have

$$(3.4.22)\qquad P_{00}(t) = 1 - F_e(t) + \sum_{n=0}^{\infty} F_e*(F*G)^{n*}*G*[1 - F](t),$$

$$(3.4.23)\qquad P_{10}(t) = \sum_{n=0}^{\infty} G_e*(F*G)^{n*}*[1 - F](t),$$

respectively. The Laplace-Stieltjes transforms of $F_e(t)$, $G_e(t)$, $F(t)$, and $G(t)$ are defined by $F_e^*(s)$, $G_e^*(s)$, $F^*(s)$, and $G^*(s)$, respectively, where

$$(3.4.24)\qquad F_e^*(s) = (\lambda/s)[1 - F^*(s)],$$

$$(3.4.25)\qquad F_e^*(s) = (\lambda/s)[1 - G^*(s)].$$

Noting that the stationary process starts from states 0 and 1 with probabilities ρ_0 and ρ_1, respectively, we have

$$(3.4.26)\qquad \rho_0 P_{00}(t) + \rho_1 P_{10}(t)$$

whose Laplace-Stieltjes transform is given by

$$(3.4.27)\qquad \rho_0 P_{00}^*(s) + \rho_1 P_{10}^*(s)$$
$$= \frac{\mu}{\lambda + \mu}\{1 - F_e^*(s) + \frac{F_e^*(s)G^*(s)[1 - F^*(s)]}{1 - F^*(s)G^*(s)}\}$$
$$+ \frac{\lambda}{\lambda + \mu} \cdot \frac{G_e^*(s)[1 - F^*(s)]}{1 - F^*(s)G^*(s)}$$
$$= \frac{\mu}{\lambda + \mu} .$$

That is,

$$(3.4.28)\qquad \rho_0 P_{00}(t) + \rho_1 P_{10}(t) = \frac{\mu}{\lambda + \mu}$$

which is independent of time t and is the stationary

transition probability in state 0. Similarly, we can show that

$$(3.4.29) \qquad \rho_0 P_{01}(t) + \rho_1 P_{11}(t) = \frac{\lambda}{\lambda + \mu}$$

by applying renewal theoretic arguments.

Let us consider a graphical representation of a Markov renewal process. Among several graphical representations, we use a signal-flow graph for representing a Markov renewal process on the Laplace-Stieltjes domains. For the alternating renewal process, we show Fig. 3.4.2, where each state in the graph corresponds to each state in Markov renewal process, and each branch gain corresponds to each element of the Laplace-Stieltjes transform of semi-Markov kernel **Q**. Note that there is no branch gains from state i to state j if $Q_{ij}(t) = 0$.

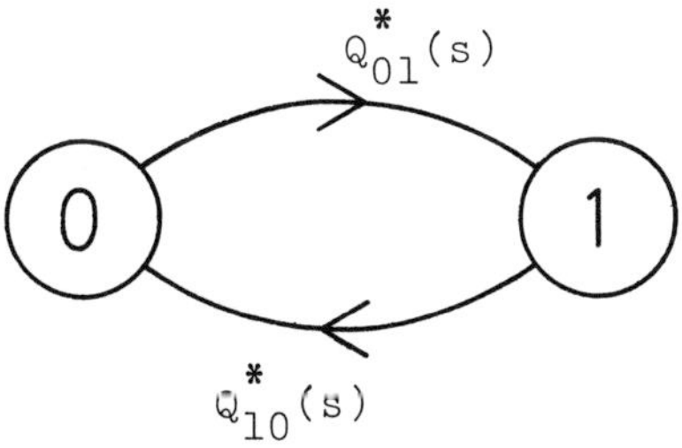

Fig. 3.4.2. A signal-flow graph representation for an alternating renewal process.

As shown in the above example of an alternating renewal process, we can easily construct a signal-flow graph for representing a Markov renewal process on the Laplace-Stieltjes transform domains. Applying the well-known Mason's gain formula for signal-flow graphs, we can derive the probabilistic quantities of interest.

Let us show how to derive the Laplace-Stieltjes transform $G^*_{ij}(s)$ (i, j = 0, 1, ..., m). Define an absorbing state m for convenience. Then the remaining states i = 0, 1, ..., m-1 are transient. It is easy to show the following:

$$(3.4.30) \quad G_{im}(t) = Q_{im}(t) + \sum_{j=0}^{m-1} Q_{ij} * G_{jm}(t).$$

The starred functions $G^*_{im}(s)$ and $Q^*_{ij}(s)$ (i, j = 0, 1, ..., m-1) denote the Laplace-Stieltjes transforms of $G_{im}(t)$ and $Q_{ij}(t)$, respectively. Introducing the mx1 vectors $\mathbf{G}^*_m(s)$ and $\mathbf{Q}^*_m(s)$ with respective component $G^*_{im}(s)$ and $Q^*_{im}(s)$ (i = 0, 1, 2, ..., m-1) and the mxm matrix $\mathbf{q}^*(s)$ with element $Q^*_{ij}(s)$ (i, j = 0, 1, 2, ..., m-1), we have the following matrix form:

$$(3.4.31) \quad \mathbf{G}^*_m(s) = \mathbf{Q}^*_m(s) + \mathbf{q}^*(s)\mathbf{G}^*_m(s)$$

or

$$(3.4.32) \quad \mathbf{G}^*_m(s) = [\mathbf{I} - \mathbf{q}^*(s)]^{-1}\mathbf{Q}^*_m(s),$$

where $\mathbf{I}$ is the mxm identity matrix. We note that the inverse matrix $[\mathbf{I} - \mathbf{q}^*(s)]^{-1}$ exists for Re(s) > 0.

Consider a Markov renewal process whose states are defined and whose semi-Markov kernel is given. Using the states defined and the Laplace-Stieltjes transform $Q^*_{ij}(s)$

(i, j = 0, 1, 2, ..., m), we can construct a state transition diagram which may be also considered to be a signal-flow graph. In the graph each node corresponds to each state and each branch gain to $Q^*_{ij}(s)$. We shall consider an algorithm for deriving $G^*_{im}(s)$ by using the signal-flow graph. As is anticipated, $G^*_{im}(s)$ can be derived by using Mason's gain formula (see Appendix B) in the signal-flow graph, where we define that node i is a source and node m is a sink. We shall verify the above fact by using the results of signal-flow graphs.

Since node i in the graph has both incoming and outcoming branches in general, we define a new source α which has an outcoming branch to node i with its branch gain a unity (see Fig. 3.4.3). Let us define the corresponding variables of nodes α, 0, 1, 2, ..., m by x_α, x_0, x_1, x_2, ..., x_m. Each branch gain corresponds to each $Q^*_{ij}(s)$ (in particular, $Q^*_{\alpha i}(s) = 1$ and $Q^*_{\alpha j}(s) = 0$ for $j \neq i$). Using the rule of signal-flow graphs, we have the following set of simultaneous linear equations:

$$
(3.4.33)\quad \begin{cases}
x_0 = Q^*_{00}(s)x_0 + Q^*_{10}(s)x_1 + \dots + Q^*_{m-1,0}(s)x_{m-1} \\
\vdots \\
x_i = Q^*_{0i}(s)x_0 + Q^*_{1i}(s)x_1 + \dots + Q^*_{m-1,i}(s)x_{m-1} \\
\qquad\qquad + x_\alpha \\
\vdots \\
x_{m-1} = Q^*_{0,m-1}(s)x_0 + Q^*_{1,m-1}(s)x_1 + \dots \\
\qquad\qquad + Q^*_{m-1,m-1}(s)x_{m-1}
\end{cases}
$$

$$
(3.4.34)\quad x_m = Q^*_{0m}(s)x_0 + Q^*_{1m}(s)x_1 + \dots + Q^*_{m-1,m}(s)x_{m-1}.
$$

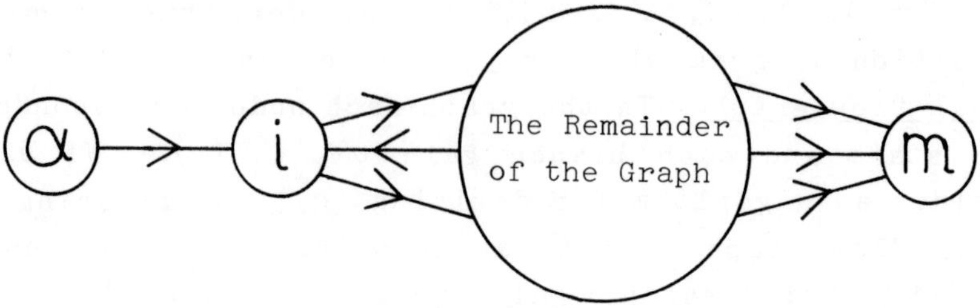

Fig. 3.4.3. A signal-flow graph with the source connected only to one node of the system.

Define the mx1 vectors

$$(3.4.35) \qquad \mathbf{x} = \begin{bmatrix} x_0 \\ \cdot \\ \cdot \\ \cdot \\ x_{m-1} \end{bmatrix}, \qquad \mathbf{e}_i = \begin{bmatrix} 0 \\ \cdot \\ \cdot \\ 1 \\ \cdot \\ \cdot \\ 0 \end{bmatrix},$$

where $\mathbf{e}_i$ is a vector with $(i+1)^{th}$ component a unity and the other components zeros. Using $\mathbf{q}^*(s)$, $\mathbf{Q}_m^*(s)$, and (3.4.35), we have from (3.4.33) and (3.4.34) the following equations:

$$(3.4.36) \qquad \mathbf{x} = \mathbf{q}^*(s)^T\mathbf{x} + \mathbf{e}_i x_\alpha \ ,$$

$$(3.4.37) \qquad x_m = \mathbf{x}\mathbf{Q}_m^*,$$

where the superscript "T" denotes the transpose of the matrix. Noting that the inverse matrix $[\mathbf{I} - \mathbf{q}^*(s)]^{-1}$ exists for $\mathrm{Re}(s) > 0$, we have the ratio x_m/x_α as follows:

$$(3.4.38) \qquad x_m/x_\alpha = \mathbf{e}_i[\mathbf{I} - \mathbf{q}^*(s)]^{-1}\mathbf{Q}_m^*(s),$$

which is the system gain assuming that node α is a source and node m is a sink. That is, the system gain coincides with $G_{im}^*(s)$ given in

$$(3.4.39) \qquad \mathbf{G}_m^*(s) = [\mathbf{I} - \mathbf{q}^*(s)]^{-1}\mathbf{Q}_m^*(s) \ .$$

As described above, deriving the Laplace-Stieltjes transform of the first passage time distribution from state i to state m in a Markov renewal process is obtaining the system gain assuming that state i is a source and state m is a sink, where the signal-flow graph is the state transition diagram in the Markov renewal process and each branch gain corresponds to each $Q_{ij}^*(s)$. To obtain the system gain in the graph, we can apply Mason's gain formula (see Appendix B), which is an easy mechanical procedure. In particular, it is more efficient to obtain the system gain for the complicated systems.

We shall derive the system reliability by using the signal-flow graph method. In the above analysis we make an additional state α which is a source. We should always consider state α . However, we omit state α and we regard state i as a source in the analysis below.

Let $M^*_{ij}(s)$ and $P^*_{ij}(s)$ denote the Laplace-Stieltjes transforms of $M_{ij}(t)$ and $P_{ij}(t)$, respectively. Then the matrices of $\mathbf{M}^*(s)$ and $\mathbf{P}^*(s)$ of $M^*_{ij}(s)$ and $P^*_{ij}(s)$, respectively, are given by

$$(3.4.40)\quad \mathbf{M}^*(s) = [\mathbf{I} - \mathbf{Q}^*(s)]^{-1}\mathbf{Q}(s),$$

$$(3.4.41)\quad \mathbf{P}^*(s) = [\mathbf{I} - \mathbf{Q}^*(s)]^{-1}[\mathbf{I} - \mathbf{H}^*(s)],$$

where $\mathbf{Q}^*(s)$ is a matrix of $Q^*_{ij}(s)$ and $\mathbf{H}^*(s)$ is a diagonal matrix of the Laplace-Stieltjes transform $H^*_i(s)$ of the unconditional distribution in state i, $H_i(t)$. Thus we can write $\mathbf{M}^*(s)$ and $\mathbf{P}^*(s)$ in the similar forms of $G^*_{im}(s)$ in (3.4.39). Modifying a method of obtaining $G^*_{im}(s)$ in (3.4.32), we can similarly obtain $\mathbf{M}^*(s)$ and $\mathbf{P}^*(s)$. In the sequel, we show how to obtain $\mathbf{M}^*(s)$ and $\mathbf{P}^*(s)$ in the signal-flow graph using a two-unit standby redundant model in reliability theory.

Consider a two-unit standby redundant model as an example and apply the signal-flow graph techniques to this model. Consider a two-unit standby redundant system of two identical units, where it is possible to repair two failed units simultaneously. Of course, the repair time of each failed unit be a random variable with arbitrary distribution $F(t)$ and the repair time of each failed unit be a random variable with exponential distribution $1 - \exp(-\mu t)$. It is assumed that a repaired unit is as good as new. It is also assumed that each switchover is perfect and each switchover time is instantaneous. It is finally assumed that a unit in standby neither deteriorates nor fails.

Define state i (i = 0, 1, 2) of the process as a number of failed units. Also define the time instants at which the process makes a transition into:

State 0: A unit begins to operate and another is in standby.

State 1: A unit begins to operate and another begins to be repaired.

State 2: An operating unit fails while another is under repair, which causes a system failure.

Since the above time instants are regeneration points, we can apply Markov renewal process to this model. The Laplace-Stieltjes transform of the mass functions (i.e., one-step transition probabilities) of a Markov renewal process under consideration are

$$Q^*_{01}(s) = \int_0^\infty e^{-st}dF(t) = F^*(s), \tag{3.4.42}$$

$$Q^*_{11}(s) = \int_0^\infty e^{-st}(1 - e^{-\mu t})dF(t)$$

$$= F^*(s) - F^*(s+\mu), \tag{3.4.43}$$

$$Q^*_{12}(s) = \int_0^\infty e^{-st}e^{-\mu t}dF(t) = F^*(s+\mu), \tag{3.4.44}$$

$$Q^*_{21}(s) = \int_0^\infty e^{-st}(2\mu)e^{-2\mu t}dt = 2\mu/(s + 2\mu). \tag{3.4.45}$$

Thus we have a signal-flow graph in Fig. 3.4.4, where each branch gain is given by $Q^*_{ij}(s)$.

Consider the Laplace-Stieltjes transform $G^*_{02}(s)$ of the first passage time distribution $G_{02}(t)$. We regard states 0 and 2 as a source and a sink, respectively, in the signal-flow graph in Fig. 3.4.4. Applying the Mason's gain formula in the graph shown in Fig. 3.4.4 with a source 0 and a sink 2, we have

$$G^*_{02}(s) = Q^*_{01}(s)Q^*_{12}(s)/[1 - Q^*_{11}(s)]. \tag{3.4.46}$$

If we regard states 1 and 2 as a source and a sink, respectively, we have the following Laplace-Stieltjes transform of the first passage time distribution $G_{12}(t)$:

$$G^*_{12}(s) = Q^*_{12}/[1 - Q^*_{11}(s)]. \tag{3.4.47}$$

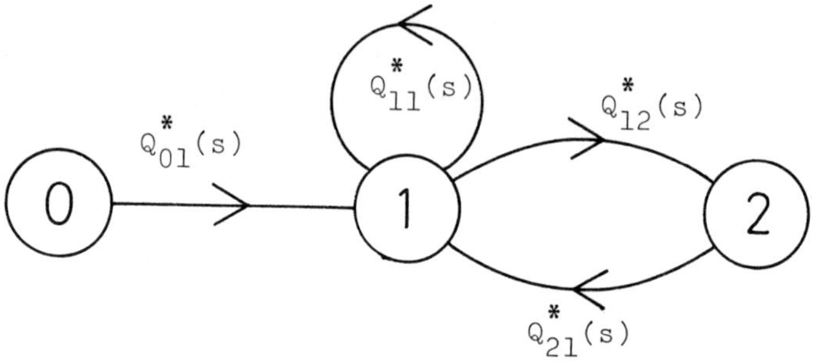

Fig. 3.4.4. A signal-flow graph of a two-unit standby redundant system.

Osaki (1970) also showed the first and second moments of the first passage time $G_{im}(t)$ by modifying the signal-flow graph. However, we omit his results. We just cite the first moments of $G_{02}(t)$ and $G_{12}(t)$:

(3.4.48) $\ell_{02} = 1/\lambda + 1/\{\lambda[1 - Q^*_{11}(0)]\}$,

(3.4.49) $\ell_{12} = 1/\{\lambda[1 - Q^*_{11}(0)]\}$,

where $1/\lambda = \int_0^\infty t dF(t)$.

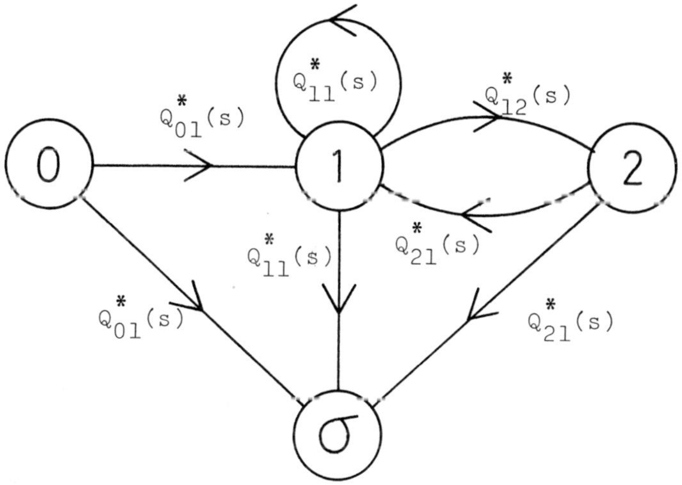

Fig. 3.4.5. A signal-flow graph for obtaining $M^*_{i1}(s)$.

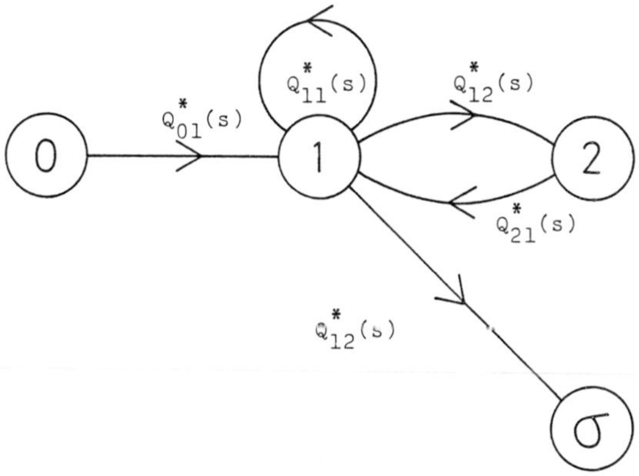

Fig. 3.4.6. A signal-flow graph for obtaining $M^*_{i2}(s)$.

Consider $G_{ij}^*(s)$, the Laplace-Stieltjes transform of the first passage time distribution from state i to state j, in general. Modifying the results above, we can extend to obtaining $G_{ij}^*(s)$ for any i and j. That is, regarding states i and j as a source and a sink, respectively, we can construct a method of obtaining $G_{ij}^*(s)$ for any i and j.

Let us next consider a method of obtaining $M_{ij}^*(s)$. Comparing (3.4.40) to (3.4.32), we can consider $\mathbf{Q}^*(s)$ instead of $\mathbf{Q}_m^*(s)$. For instance, $M_{i1}^*(s)$ (i = 0, 1, 2) can be obtained from that state i is a source and a new sink σ is added with branch gain $Q_{i\sigma}^*(s) = Q_{i1}^*(s)$ (i = 0, 1, 2, 3), which is illustrated in Fig. 3.4.5. Further, from Fig. 3.4.6, we can obtain $M_{i2}^*(s)$ (i = 0, 1, 2). Thus we have

$$(3.4.50)\quad \mathbf{M}^*(s) = [1 - Q_{11}^*(s) - Q_{12}^*(s)Q_{21}^*(s)]^{-1}$$

$$\times \begin{bmatrix} 0 & Q_{01}^*(s) & Q_{01}^*(s)Q_{12}^*(s) \\ 0 & Q_{11}^*(s) + Q_{12}^*(s)Q_{21}^*(s) & Q_{12}^*(s) \\ 0 & Q_{21}^*(s) & Q_{21}^*(s)Q_{12}^*(s) \end{bmatrix}.$$

Let us finally consider $P_{ij}^*(s)$. Since the matrix $\mathbf{P}^*(s)$ of $P_{ij}^*(s)$ is just a similar form of (3.4.41), we can construct a similar method of obtaining $P_{ij}^*(s)$. For instance, $P_{i1}^*(s)$ (i = 0, 1, 2) can be obtained by adding a new sink σ with branch gains $Q_{1\sigma}^*(s) = 1 - H_i^*(s)$ and defining a new source i (i = 0, 1, 2) (see Fig. 3.4.7). Thus we have

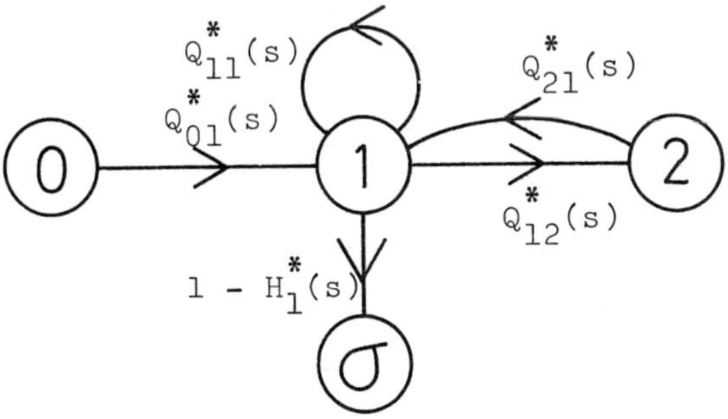

(i) For obtaining $P^*_{i1}(s)$ $(i = 0, 1, 2)$

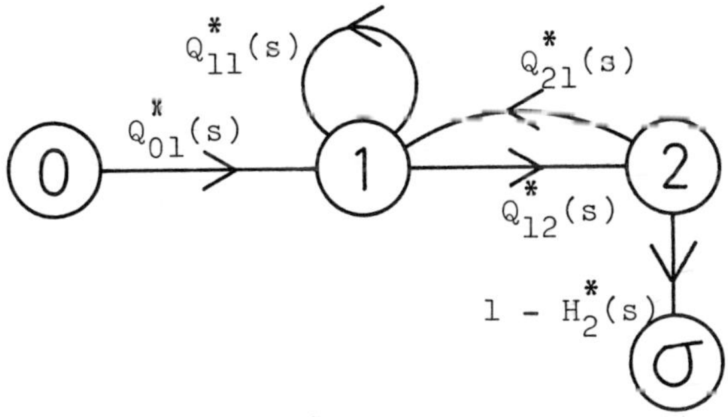

(ii) For obtaining $P^*_{i2}(s)$ $(i = 0, 1, 2)$

Fig. 3.4.7. A signal-flow graph for obtaining $P^*_{ij}(s)$.

$$(3.4.51)\quad \mathbf{P}^*(s) = [1 - Q^*_{11}(s) - Q^*_{12}(s)Q^*_{21}(s)]^{-1}$$

$$\times \begin{bmatrix} 0 & Q^*_{01}(s)[1 - H^*_1(s)] & Q^*_{01}(s)Q^*_{12}(s)[1 - H^*_2(s)] \\ 0 & [1 - H^*_1(s)] & Q^*_{12}(s)[1 - H^*_2(s)] \\ 0 & Q^*_{21}(s)[1 - H^*_1(s)] & [1 - H^*_2(s)] \end{bmatrix}.$$

If we assume that the mean time to failure and mean repair time of each unit are finite, then there exist the limiting probabilities $P_j = \lim_{t\to\infty} P_{ij}(t)$ (i, j = 0, 1, 2) which are independent of an initial state i. That is, denoting

(3.4.52) $$1/\lambda \equiv \int_0^\infty t dF(t) ,$$

we have

(3.4.53) $$P_0 = 0 ,$$

(3.4.54) $$P_1 = (1/\lambda)/[1/\lambda + F^*(\mu)/(2\mu)] ,$$

(3.4.55) $$P_2 = [F^*(\mu)/(2\mu)]/[1/\lambda + F^*(\mu)/(2\mu)] .$$

As state 0 is a transient state in this model, it is evident that $G_{i0}^*(s)$, $M_{i0}^*(s)$, and $P_{i0}^*(s)$ (i = 0, 1, 2) are equal to zeros.

We have discussed the relationship between Markov renewal processes and signal-flow graphs, and shown a method of obtaining $\mathbf{G}^*(s)$, $\mathbf{M}^*(s)$, and $\mathbf{P}^*(s)$, by modifying the corresponding signal-flow graphs.

It is tedious to calculate $G_{ij}^*(s)$, $M_{ij}^*(s)$, and $P_{ij}^*(s)$ directly from the results of Markov renewal processes. We need matrix inversion techniques to do so. However, a method of signal-flow graphs are easier and can be done by Mason's gain formula which is accessible for any one who has no knowledge about linear algebra. In particular, if we do not need all the quantities above but need some specified ones, a method of signal-flow graphs is convenient to do so without cumbersome matrix inversion.

3.5 Markov Renewal Processes with Non-Regeneration Points

We have discussed some models in reliability theory by investigating the relationship between Markov renewal processes and signal-flow graphs. A Markov process or semi-Markov process is one of the most powerful mathematical techniques for analyzing stochastic models in reliability theory. However, it is very complicated to investigate some state of the process having non-regeneration points, i.e., some time instants at which the process moves some states which are not regeneration points. For example, if we are interested in the availability for a two-unit standby redundant system, we have to consider the time instants at which system failure occurs, which is no longer a regeneration point under some conditions (see Barlow and Proschan (1975), p.203. Then we show how to analyze such situations in detail. Of course, such a two-unit standby redundant system can be analyzed as shown in the sequel.

This section proposes unique modifications of the Markov renewal processes and considers applications of such Markov renewal processes to analysis of redundant repairable systems including some non-regeneration points. We define new mass functions and derive (i) the first passage time distributions, (ii) the transition probabilities, and (iii) the renewal functions, with the aids of renewal-theoretic arguments.

We consider a Markov renewal process $\{\mathbf{N}(t),\ t \geq 0\}$, where the state is denoted by integers $0, 1, 2, \ldots, n \in S$ and the state space is finite. Recall that $N_i(t)$ denote the number of visits to state i during the interval $[0, t]$.

Recall

$$(3.5.1) \qquad P_{ij}(t) = P\{Z(t) = j | Z(0) = i\},$$

which denotes the probability that the process is in state j at time t if the process starts in state i at time 0. Further, we recall that

$$(3.5.2) \qquad G_{ij}(t) = P\{N_j(t) > 0 | Z(0) = i\},$$

and

$$(3.5.3) \qquad M_{ij}(t) = E[N_j(t) | Z(0) = i].$$

Throughout this section we assume that the Markov renewal process under consideration has only one (positive) recurrent class, because we restrict ourselves to applications to reliability theory. Assume also that each $G_{jj}(t)$ is not lattice for any state j. It follows from the previous result that

$$(3.5.3) \qquad \ell_{jj} = \int_0^\infty t dG_{jj}(t)$$

is finite, and that there exist both the limiting probabilities

$$(3.5.4) \qquad P_j = \lim_{t\to\infty} P_{ij}(t)$$

and

$$(3.5.5) \qquad M_j = \lim_{t\to\infty} M_{ij}(t),$$

which are independent of the initial state i.

Consider a case where time instants at which the process move some states are not regeneration points. Then, we partition the state space S into $S = S^* \cup S^+$ ($S^* \cap S^+ = \emptyset$), where S^* is the portion of the state space such that the time instants moving state i ($i \in S^*$) are not regeneration points, and S^+ is the remainder of the state space such that the time instants moving state i ($i \in S^+$) are regeneration points. Note that we assume that S^* is not the empty set.

Define the mass function $Q_{ij}(t)$ from state i ($i \in S^+$) to state j ($j \in S$) by the probability that after moving state i, the process makes a transition into state j, in an amount of time less than or equal to time t. However, it is impossible to define mass functions $Q_{ij}(t)$ if $i \in S^*$, since the time instant moving state i is not a regeneration point. Define new mass functions $Q_{ij}^{(k_1,k_2,\ldots,k_m)}(t)$ which is the probability that after moving state i ($i \in S^+$), the process next makes transitions into states $k_1, k_2, \ldots, k_m$ ($k_1, k_2, \ldots, k_m \in S^*$) and finally moves state j ($j \in S$), in an amount of time less than or equal to time t. Moreover, we define

$$(3.5.7) \qquad H_i(t) = \sum_{j \in S} Q_{ij}(t) \qquad (i \in S^+),$$

which is the unconditional distribution of the time elapsed from state i to the next state moved (possibly i itself).

Type 1 - Markov Renewal Process

Consider a Markov renewal process with $n + 1$ states, which consists of $S^+ = \{0\}$ and $S^* = \{1, 2, ..., n\}$ (see Fig. 3.5.1). The process starts from state 0, makes transitions into states 1, 2, ..., n, and comes back to state 0. Suppose that $Z(0) = 0$. Then, from straightforward renewal-theoretic arguments, the first-passage time distributions are

$$(3.5.8) \quad G_{01}(t) = Q_{01}(t) \ ,$$

$$(3.5.9) \quad G_{0j}(t) = Q_{0j}^{(1,2,\dots,j-1)}(t) \qquad (j = 2, 3, \dots, n),$$

$$(3.5.10) \quad G_{00}(t) = Q_{00}^{(1,2,\dots,n)}(t).$$

The renewal functions are

$$(3.5.11) \quad M_{01}(t) = Q_{01}(t) + Q_{00}^{(1,2,\dots,n)} * M_{01}(t)$$

$$= Q_{01}(t) + G_{00} * M_{01}(t),$$

$$(3.5.12) \quad M_{0j}(t) = Q_{0j}^{(1,2,\dots,j-1)}(t) + G_{00} * M_{0j}(t) \qquad (j - 2, 3, \dots, n),$$

$$(3.5.13) \quad M_{00}(t) = G_{00}(t) + G_{00} * M_{00}(t).$$

The transition probabilities are

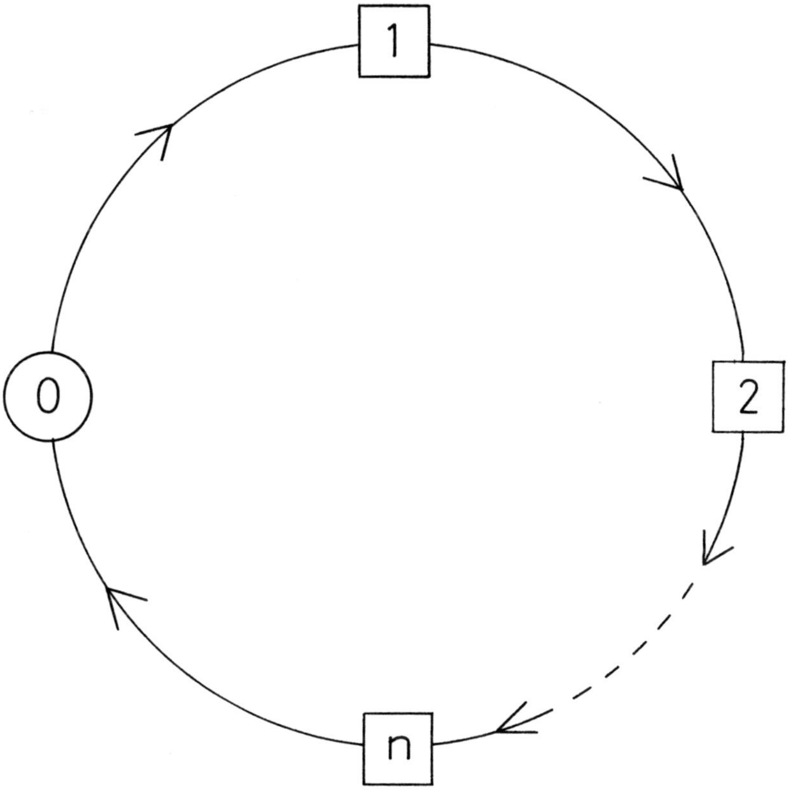

Fig. 3.5.1. The state transition diagram for the Type 1 - Markov renewal process, where ○ represents a state with regeneration point and □ a state with non-regeneration point.

$$(3.5.14)\quad P_{01}(t) = Q_{01}(t) - Q_{02}^{(1)}(t) + Q_{00}^{(1,2,\ldots,n)}*P_{01}(t)$$

$$= Q_{01}(t) - Q_{02}^{(1)}(t) + G_{00}*P_{01}(t),$$

$$(3.5.15)\quad P_{0j}(t) = Q_{0j}^{(1,2,\ldots,j-1)}(t) - Q_{0,j+1}^{(1,2,\ldots,j)}(t)$$

$$+ G_{00}*P_{0j}(t) \qquad (j = 2, 3, \ldots, n),$$

$$(3.5.16)\quad P_{00}(t) = 1 - Q_{01}(t) + G_{00}*P_{00}(t),$$

where $Q_{0,n+1}^{(1,2,\ldots,n)}(t) = Q_{00}^{(1,2,\ldots,n)}(t)$ and the asterisk denotes the Stieltjes convolution. Taking the Laplace-Stieltjes transforms on both sides of (3.5.11) - (3.5.16), we can obtain the Laplace-Stieltjes transforms of $G_{0j}(t)$, $M_{0j}(t)$, and $P_{0j}(t)$ $(j = 0, 1, 2, \ldots, n)$. It is evident that

$$(3.5.17)\quad \sum_{j=0}^{n} P_{0j}(t) = 1.$$

From the Laplace-Stieltjes transforms, the renewal functions and the transition probabilities can be obtained explicitly upon inversion. However, it is not easy except the simplest cases.

((3.5.18) Example Consider an $(n + 1)$-unit parallel redundant system. If at least one of the $n + 1$ units is operating, then the system is operating. If all the units are down simultaneously, then the system fails and will begin to operate again immediately by replacing all the failed units by the new ones. All units operate independently and each unit has an identical failure time distribution $F(t)$ which is non-lattice and has a finite mean $1/\lambda$. We denote the states by the total number of the failed units. Suppose that all the units begins to operate at time 0. Then the mass functions are

(3.5.19) $$Q_{01}(t) = 1 - [\bar{F}(t)]^{n+1},$$

(3.5.20) $$Q_{0j}^{(1,2,\ldots,j-1)}(t) = \sum_{k=j}^{n+1} \binom{n+1}{k} [F(t)]^k [\bar{F}(t)]^{n+1-k},$$

where $\bar{F}(t) \equiv 1 - F(t)$. Thus, substituting the above equations into (3.5.11) - (3.5.16), we can obtain the renewal functions and the transitions probabilities.

Type 2 - Markov Renewal Process

Consider a Markov renewal process consisting of $S^+ = \{0\}$ and $S^* = \{1, 2, \ldots, n\}$ (see Fig. 3.5.2). The process starts from state 0 and is permitted only to make a transition into one state j ($j \varepsilon S^*$). Then the process returns to state 0, and so on. For instance, the process repeats among 0, j_1, 0, j_2, ..., where $j_1, j_2, \ldots \varepsilon S^*$. The Laplace-Stieltjes transforms of the first passage time distributions, the renewal functions, and the transition probabilities are

(3.5.21) $$G_{00}^*(s) = \sum_{i=0}^{n} Q_{00}^{(i)*}(s),$$

(3.5.22) $$G_{0j}^*(s) = Q_{0j}^*(s)/[1 - G_{00}^*(s)] \quad (j = 1, 2, \ldots, n),$$

(3.5.23) $$M_{00}^*(s) = G_{00}^*(s)/[1 - G_{00}^*(s)],$$

(3.5.24) $$M_{0j}^*(s) = G_{0j}^*(s)/[1 - G_{00}^*(s)]$$

(j = 1, 2, ..., n),

(3.5.25) $$P_{00}^{*}(s) = [1 - \sum_{i=1}^{n} Q_{0i}^{*}(s)]/[1 - G_{00}^{*}(s)],$$

(3.5.26) $$P_{0j}^{*}(s) = [Q_{0j}^{*}(s) - Q_{00}^{(j)*}(s)]/[1 - G_{00}^{*}(s)]$$
(j = 1, 2, ..., n).

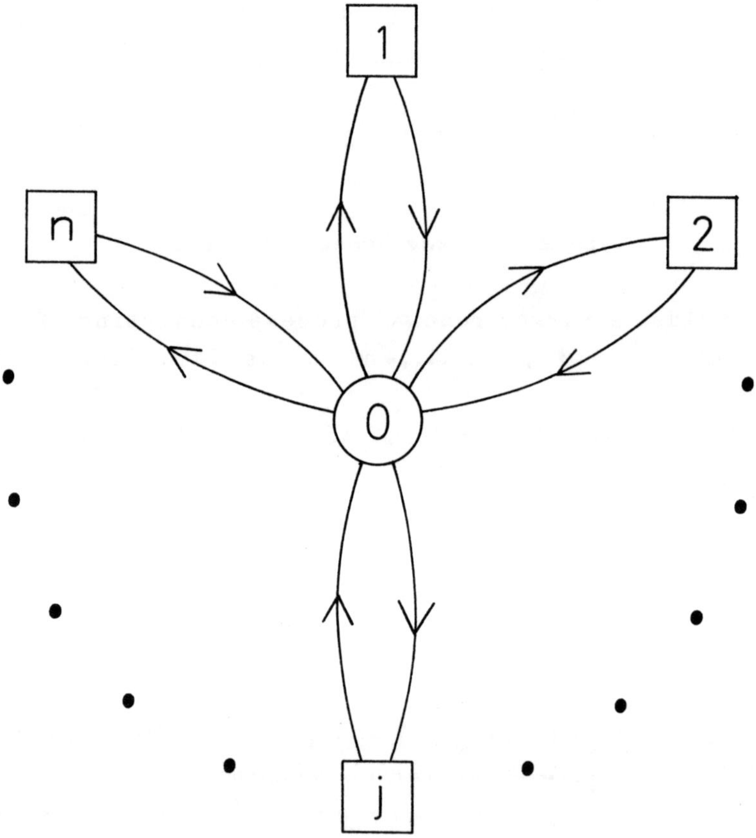

Fig. 3.5.2. The state-transition diagram for the Type 2 - Markov renewal process.

If $n = 1$ then the process corresponds to a special case of Type 1. That is, it is the simplest state space with a single non-regeneration point. The process takes two alternate states 0 and 1. If the time instant moving state 1 is also a regeneration point, then the process becomes an alternating renewal process studied in the preceding section.

(3.5.27) Example Consider a two-unit standby redundant system of two identical units. The detailed discussions will be found in Section 5.3. Define the states 0 (one unit is operating and the other unit is in standby) 1 (one unit is operating and the other unit is under repair) and 2 (one unit is under repair and the other unit waits for repair). Then, the system generates a Markov renewal process with the state space $\{0, 1, 2\}$, where $S^+ = \{1\}$ and $S^* = \{0, 2\}$. Then, the mass functions are $Q_{10}(t)$, $Q_{12}(t)$, $Q_{11}^{(0)}(t)$, and $Q_{11}^{(2)}(t)$. Thus, we can obtain the interesting quantities of the system by using the results for Type 2 - Markov renewal process. The fruitful results will be shown in Section 5.3.

Remarks and An Example

We have considered the unique modifications of the regeneration point techniques in the Markov renewal processes with some non-regeneration points. That is, the Laplace-Stieltjes transforms of the first-passage time distributions, the renewal functions, and the transition probabilities are given in terms of the mass functions for such Markov renewal processes.

We have considered two examples of the processes to redundant systems. In a similar fashion, many redundant systems can be also described by Type 1 - or Type 2 - Markov renewal processes, and the the conventional Markov renewal processes. For instance, some two-unit standby redundant systems with repair and preventive maintenance will be discussed by combining Type 1 - Markov renewal process and the conventional Markov renewal process. Note that if either the repair time distribution or the failure time distribution is not exponential, such systems have non-regeneration points.

(3.5.28) Example A model of spare parts inventory was discussed by Wiggins (1968). A unit begins to operate at time 0. If the operating unit fails before time T, the spare part is ordered immediately and is placed into service as soon as it is delivered after a lead time L. If the operating unit does not fail up to time T, the spare part is ordered at time T and is delivered after the lead time. In this case, the spare part is put into service if the original unit has failed, or is put into the spare part inventory if the original unit has not failed.

Define the states 0 (the unit is operating and the spare part is not delivered), 1 (the unit is operating and the spare part is ordered), 2 (the unit is operating and the spare part is in inventory), 3 (the operating unit has failed during the order of the spare part), and 4 (the spare part is ordered because the operating unit has failed before T).

Assume that the failure time distribution for each unit is $F(t)$. Further assume that $P\{T \leq t\} = A(t)$ and $P\{L \leq$

t} = B(t), which might be the degenerate distribution placing unit masses at t_0 and ℓ_0, respectively. Then, the mass functions of the above process (see Fig. 3.5.3) are

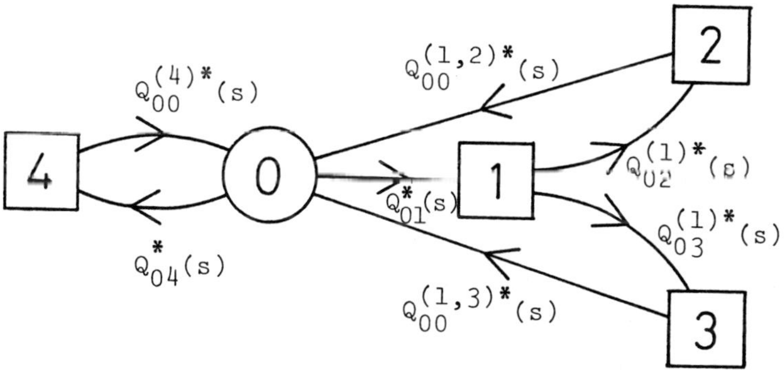

Fig. 3.5.3. The state-transition diagram for the spare parts inventory model.

$$(3.5.29)\quad Q_{01}(t) = \int_0^t \bar{F}(t)dA(t) ,$$

$$(3.5.30)\quad Q_{02}^{(1)}(t) = \int_0^t \bar{F}(u)d[A*B(u)] ,$$

$$(3.5.31)\quad Q_{00}^{(1,2)}(t) = \int_0^t [A*B(u)]dF(u) ,$$

$$(3.5.32)\quad Q_{03}^{(1)}(t) = \int_0^t [A*\bar{B}(u)]dF(u) ,$$

$$(3.5.33)\quad Q_{00}^{(1,3)}(t) = \int_0^t \int_0^u [F(u) - F(v)]dA(v)dB(u-v) ,$$

$$(3.5.34)\quad Q_{04}(t) = \int_0^t \bar{A}(u)dF(u) ,$$

$$(3.5.35)\quad Q_{40}(t) = B(t).$$

In particular, the Laplace-Stieltjes transforms of the transition probabilities are

$$(3.5.36)\quad P_{00}^*(s) = [1 - Q_{01}^*(s) - Q_{04}^*(s)]/[1 - H_{00}^*(s)],$$

$$(3.5.37)\quad P_{01}^*(s) = [Q_{01}^*(s)-Q_{02}^{(1)*}(s)-Q_{03}^{(1)*}(s)]/[1 - H_{00}^*(s)],$$

$$(3.5.38)\quad P_{02}^*(s) = [Q_{02}^{(1)*}(s) - Q_{00}^{(1,2)*}(s)]/[1 - H_{00}^*(s)],$$

$$(3.5.39)\quad P_{03}^*(s) = [Q_{03}^{(1)*}(s) - Q_{00}^{(1,3)*}(s)]/[1 - H_{00}^*(s)],$$

$$(3.5.40)\quad P_{04}^*(s) = Q_{04}^*(s)[1 - Q_{40}^*(s)]/[1 - H_{00}^*(s)],$$

where

$$(3.5.41)\quad H_{00}^*(s) = Q_{00}^{(1,2)*}(s) + Q_{00}^{(1,3)*}(s) + Q_{04}^*(s)Q_{40}^*(s),$$

which is the recurrence time distribution for state 0. In a similar way, we can obtain the first passage time distributions and the renewal functions.

We believe that the unique modifications for the regeneration point techniques discussed in this section will be frequently used in the sequel discussions and can be applied in other contexts, e.g., inventory control and queueing theory.

Bibliography and Comments

Throughout all the sections: Markov renewal process (or semi-Markov processes) were summarized by Pyke (1961a, 1961b) and Cinlar (1975). We are following the basic ideas of Cinlar (1975).

Section 3.3: Barlow and Proschan (1965) showed the first and second moments of the first passage time distribution. Pyke (1961b) showed the stationary Markov renewal processes.

Section 3.4: The relationship between Markov renewal processes and signal-flow graphs was discussed by Osaki (1970, 1974).

Section 3.5: We are following the papers by Nakagawa and Osaki (1974, 1976). Osaki and Nishio (1980) showed the procedure of analysis for Markov renewal processes with non-regeneration points, and applied such the procedure to reliability models with many states. A two-unit standby redundant system was thoroughly analyzed by Nakagawa and Osaki (1974). Barlow and Proschan (1975) showed the limiting probabilities for such a two-unit standby redundant system by applying the heuristic approach. Wiggins (1968) discussed a spare parts inventory model which was formulated by a Markov renewal process with non-regeneration points.

[1] R.E. Barlow and F. Proschan (1965), Mathematical Theory of Reliability, Wiley, New York.

[2] R.E. Barlow and F. Proschan (1975), Statistical Theory of Reliability and Life Testing - Probability Models, Holt, Rinehart, and Winston, New York.

[3] E. Cinlar (1975), Introduction to Stochastic Processes, Prentice-Hall, Englewood Cliffs, N.J.

[4] T. Nakagawa and S. Osaki (1974), "Stochastic Behaviour of a Two-Unit Standby Redundant System," INFOR, Vol. 12, pp. 66-70.

[5] T. Nakagawa and S. Osaki (1976), "Markov Renewal Processes with Some Non-Regeneration Points and Their Applications to Reliability Theory," Microelectron. Reliab., Vol. 15, pp. 633-636.

[6] S. Osaki (1970), "System Reliability Analysis by Markov Renewal Processes," J. Operations Res. Soc. Japan, Vol. 12, pp. 127-188.

[7] S. Osaki (1974), "Signal-Flow Graphs in Reliability Theory," Microelectron. Reliab., Vol. 13, pp. 539-541.

[8] S. Osaki and T. Nishio (1980), Reliability Evaluation of Some Fault-Tolerant Computer Architectures, Springer-Verlag, Berlin.

[9] R. Pyke (1961a), "Markov Renewal Processes: Definitions and Preliminaries," Ann. Math. Statist., Vol. 32, pp. 1231-1242.

[10] R. Pyke (1961b), "Markov Renewal Processes with Finitely Many States," Ann. Math. Statist., Vol. 32, pp. 1243-1259.

[11] A.D. Wiggins (1967), "A Minimum Cost Model of Spare Parts Inventory Control," Technometrics, Vol. 9, pp. 661-665.

CHAPTER 4

STOCHASTIC MODELS FOR ONE-UNIT SYSTEMS

4.1 Introduction

In the preceding two chapters we have discusssed stochastic processes which are of direct use to analyze stochastic models in reliability theory. In this chapter we are concerned with stochastic models for one-unit systems which are basic in system reliability modeling.

In Section 4.2 we develop the availability theory for one-unit systems which assume up and down states alternately. Several availability measures will be derived by applying Markov renewal theory of special type (i.e., alternating renewal processes, see Section 3.4).

In Section 4.3 we study several basic replacement models and derive optimal replacement models under suitable assumptions and criteria. Such replacement models are age replacement models, block replacement models, etc. We also discuss discrete models and discounted models for such replacement models.

In Section 4.4 we develop ordering policies in which each spare can be provided only by an order after a lead time in advance. Optimal ordering policies are obtained by minimizing the expected cost per unit time in the steady-state. Discounted and discrete models are also discussed.

In Section 4.5 we develop inspection policies in which failure can be detected only by inspections. We also discuss optimal inspection policies which minimize the total expected cost up to failure detection. Several variations of such inspection policies will be also discussed.

In the remaining of this section we summarize the terms of reliability, and several reliability and availability measures for systems. Such measures will be frequently used for later discussions.

First of all, we introduce the following basic terms from MIL-STD 721B.

(4.1.1) Definition Item: A non-specific term used to denote any product, including systems, materials, parts, subassembles, sets, accessaries, etc.

In this book we use systems and items interchangeably. However, in the above definition, an "item" includes a system. We prefer a system to an item in this book.

(4.1.2) Definition Failure: The event, or inoperable state, in which any item or part of an item does not, or would not, perform as previously specified.

In particular, we are interested in random failure.

(4.1.3) Definition Random Failure: Failure whose occurrence is predictable only in a probabilistic or statistical sense. This applies to all distributions.

Let X denote the lifetime of a system or an item subject to random failure, which is, of course, a random variable. The distribution of the lifetime to failure is given by

$$F(t) = P\{X \leq t\} \qquad (t \geq 0). \tag{4.1.4}$$

In this chapter we assume that the failure law of a system is known, i.e., the lifetime distribution $F(t)$ in (4.1.4) is known.

In Section 4.2 and the following chapters we are very much interested in the behavior after failure. That is, a failed item is maintained or repaired. The following terms are related to the terms of maintenance from MIL-STD 721B.

(4.1.5) Definition Maintenance: All actions necessary for retaining an item in or restoring it to a specified condition.

(4.1.6) Definition Maintainability: The measure of the ability of an item to be retained in or restored to specific condition when maintenance is performed by personnel having specified skill levels, using prescribed procedures and resources, at prescribed level of maintenance and repair.

(4.1.7) **Definition** Corrective Maintenance: All actions performed as a result of failure, to restore an item to a specified condition. Corrective maintenance can include any or all of the following steps: Localization, Isolation, Disassembly, Interchange, Reassembly, Alignment and Checkout.

(4.1.8) **Definition** Preventive Maintenance: All actions performed in an attempt to retain an item in specified condition by providing systematic inspection, detection, and prevention of incipient failures.

(4.1.9) **Definition** Scheduled Maintenance: Preventive maintenance performed at prescribed points in the item's life.

In this chapter we are concerned with scheduled maintenance and show optimal scheduled maintenance policies minimizing or maximizing the prespecified criteria.

The following terms are from IEC.

(4.1.10) **Definition** Up Time: The period of time during which an item performs its required function.

(4.1.11) **Definition** Down Time: The period of time during which an item is not in a condition to perform its required function.

Let X_i and Y_i $(i = 1, 2, \ldots)$ donote the lifetime and maintenance time of a system, where we assume that each system is identical and maintenance of the failed system can perform its required function. Consider a sample function shown in Fig. 4.1.1 that the system assumes up and down states alternately. We further assume that

(4.1.12) $F(t) = P\{X_i \leq t\}$ $(t \geq 0, i = 1, 2, \ldots)$,

(4.1.13) $G(t) = P\{Y_i \leq t\}$ $(t \geq 0, i = 1, 2, \ldots)$,

where all the random variables are assumed to be independent. Referring to Fig. 4.1.1, we are now ready to define the following reliability and availability measures.

(4.1.14) Definition Reliability $R(t)$: The probability of an item performing its required function for the intended period of time $[0, t]$.

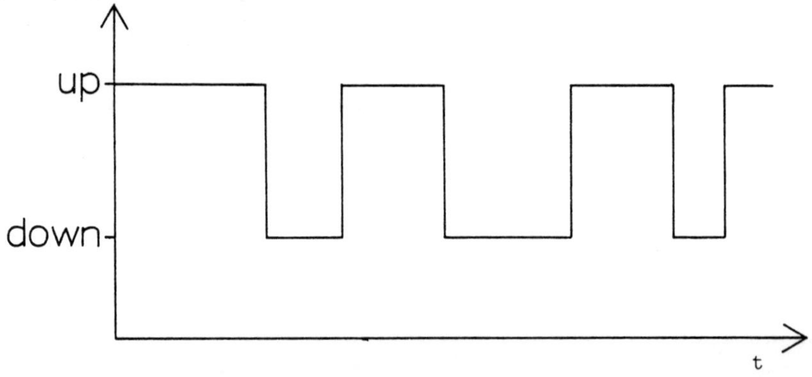

Fig. 4.1.1. A sample function of a process assuming up and down states alternately.

It is clear from Fig. 4.1.1 that the reliability is

(4.1.15) $R(t) = 1 - F(t) \quad (t \geq 0)$,

and the Mean-Time-To-Failure (MTTF) is

(4.1.16) $\mathrm{MTTF} = \int_0^\infty t\,dF(t) = \int_0^\infty R(t)\,dt$,

when $R(0) = 1$.

(4.1.17) Definition Interval Reliability $R(x,t)$: The probability that at a specified time t, a system is operating and will continue to operate for an interval of duration x.

(4.1.18) Definition Limiting Interval Reliability $R(x)$: It is defined by

(4.1.19) $R(x) = \lim_{t\to\infty} R(x,t)$.

(4.1.20) Definition Availability $A(t)$: The probability that a system is operating at a specified time t. This availability is referred to as the pointwise availability or instantaneous availability.

(4.1.21) Definition (1) Limiting Availability:

(4.1.22) $A = \lim_{t\to\infty} A(t)$

when it exists.

(2) Average Availability in $[0, T]$:

$$(4.1.23) \qquad A_{av}(T) = (1/T)\int_0^T A(t)dt.$$

(3) Limiting Average Availability:

$$(4.1.24) \qquad A_{av}(\infty) = \lim_{T\to\infty} (1/T)\int_0^T A(t)dt.$$

As will shown in Section 4.2,

$$(4.1.25) \qquad A = MUT/(MUT + MDT),$$

where the Mean Up Time (MUT) and Mean Down Time (MDT) are

$$(4.1.26) \qquad MUT = \int_0^\infty tdF(t)$$

and

$$(4.1.27) \qquad MDT = \int_0^\infty tdG(t),$$

respectively (see Fig. 4.1.1). The MUT and MDT are also referred to as the Mean-Time-Between-Failures (MTBF) and the Mean-Time-Between-Maintenances (MTBM), respectively (see, e.g., Barlow and Proschan (1975)). It is well-known that the limiting average availability is the same as limiting availability. That is, if $A = \lim_{t\to\infty} A(t)$ exists, then

$$(4.1.28) \qquad A_{av}(\infty) = \lim_{T\to\infty} (1/T)\int_0^T A(t)dt = A.$$

The following joint availability will be discussed in Section 4.2.

(4.1.29) Definition The joint availability $A_{joint}(x,t)$ at t and $t+x$: The probability that the system is operating at t and again $t+x$. The limiting joint availability is defined by

$$(4.1.30) \qquad A_{joint}(x) = \lim_{t\to\infty} A_{joint}(x,t).$$

4.2 Availability Theory

In this section we develop the availability theory for a one-unit system which assumes up and down states alternately (see Fig. 4.1.1). We assume that the failure (up) time distribution is a general $F(t)$ with finite mean $1/\lambda$ and the repair (down) time distribution is also a general $G(t)$ with finite mean $1/\mu$. We also assume that repair of the failed unit recovers its functioning perfectly. We are concerned with the stochastic behavior for such a system in an infinite time interval.

As shown in Section 3.4, the model in which we are interested can be formulated by an alternating renewal process which is a Markov renewal process of special type. Let $i = 0, 1$ denote the down and up states, respectively. Let $I_k(t)$ denote the binary indicator variable which assumes the value 1 (0) if the system is up (down), at time t, given that it was in state k at $t = 0$, respectively.

As shown in Fig. 4.1.1, the system repeats up and down states infinitely, and we define

$$(4.2.1) \qquad H(t) = F*G(t)$$

which is the distribution of the sum of each up and down times. Defining the renewal function for the underlying distribution $H(t)$, we have

$$(4.2.2) \qquad M_H(t) = \sum_{n=1}^{\infty} H^{(n)}(t)$$

which will play a central rule for later analysis. Note that from Elementary Renewal Theorem (2.3.20),

$$(4.2.3) \qquad \lim_{t\to\infty} M_H(t)/t = 1/(1/\lambda + 1/\mu) = \lambda\mu/(\lambda + \mu) \ .$$

Let $A_k(t) = P\{I_k(t) = 1\}$ $(k = 0, 1)$ denote the pointwise availability of the system given that the system was in state k at time 0, respectively. Referring to the results in Section 3.4, we have

$$(4.2.4) \qquad A_k(t) = P_{1-k,0}(t) \qquad (k = 0, 1),$$

since we have to notice that the definitions of states 0 and 1 for an alternating renewal process in Section 3.4 are just opposite to those in this section, or

$$(4.2.5) \qquad A_1(t) = \bar{F}(t) + M_H*\bar{F}(t),$$

$$(4.2.6) \qquad A_0(t) = G*\bar{F}(t) + G*M_H*\bar{F}(t).$$

It is easy to show that

$$(4.2.7) \qquad A = \lim_{t\to\infty} A_k(t) = \mu/(\lambda + \mu) \qquad (k = 0, 1),$$

which is the <u>limiting availability</u>, independent of the initial state k. Equation (4.2.7) can be rewritten by equation (4.1.25).

We are interested in the interval availability

(4.2.8) $$R_k(x,t) = P\{I_k(u) = 1,\ u \in [t,t+x] \mid X(0) = k\},$$

which is the survival probability that the system is up at time t and will continue to be up for an interval of duration x. Just similar to deriving the residual lifetime distribution in (2.3.55), we have

(4.2.9) $$R_1(x,t) = \bar{F}(t+x) + \int_0^t \bar{F}(t+x-u)dM_H(u)$$

and

(4.2.10) $$R_0(x,t)$$

$$= \int_0^t \bar{F}(t+x-u)dG(u) + \int_0^t \bar{F}(t+x-u)d[G*M_H(u)].$$

Applying the Key Renewal Theorem (2.3.48), we have

(4.2.11) $$R(x) = \lim_{t\to\infty} R_1(x,t)$$

$$= \lim_{t\to\infty} R_0(x,t) = \frac{\lambda\mu}{\lambda + \mu}\int_x^\infty \bar{F}(u)du ,$$

which is the limiting interval availability. Note that

(4.2.12) $$R(x) = \frac{\mu}{\lambda + \mu}\int_x^\infty [\bar{F}(u)/(1/\lambda)]du .$$

Let

(4.2.13) $$\psi(t) = F(t)/(1/\lambda), \quad \Psi(t) = \int_0^t \psi(u)du$$

be the density and distribution of the asymptotic distribution of the excess time (see equation (2.3.62)).

Then we have

$$(4.2.14)\qquad R(x) = \frac{\mu}{\lambda + \mu}[1 - \Psi(x)] ,$$

the right-hand side of which is the product of the limiting probability (availability) that the system is up in the steady-state and the limiting probability that it can survive an interval of duration at least x.

Let us introduce the random variable $\gamma_k(t)$, the up excess random variable to down at time t, given $I_k(t) = 1$ $(k = 0, 1)$. It is easy to show that

$$(4.2.15)\qquad P\{\gamma_k(t) > x\} = R_k(x,t)/A_k(t) \qquad (k = 0, 1),$$

and the up excess time distribution is given by

$$(4.2.16)\qquad P\{\gamma_k(t) \le x\} = 1 - R_k(x,t)/A_k(t) \quad (k = 0, 1).$$

We are also interested in the mean up excess time:

$$(4.2.17)\qquad E[\gamma_k(t)] = \int_0^\infty [R_k(x,t)/A_k(t)]dx \qquad (k = 0, 1).$$

In particular,

$$(4.2.18)\qquad E[\gamma_1(t)] = \frac{(1/\lambda + 1/\mu)}{A_1(t)}[R(t) + R*M_H(t)],$$

$$(4.2.19)\qquad E[\gamma_0(t)] = \frac{(1/\lambda + 1/\mu)}{A_0(t)}R*[G(t) + G*M_H(t)].$$

Noting that $\lim_{t\to\infty} A_k(t) = A$ $(k = 0, 1)$ and $\lim_{t\to\infty} R_k(x,t) = A[1 - \Psi(x)]$, we have

$$(4.2.20)\qquad \lim_{t\to\infty} P\{\gamma_k(t) \le x\} = \Psi(x) = \int_0^x [\bar{F}(u)/(1/\lambda)]du.$$

Assuming $t \to \infty$ in $E[\gamma_k(t)]$ $(k = 0, 1)$ and applying the Key Renewal Theorem (2.3.48), we have

(4.2.21) $$\lim_{t\to\infty} E_k(t) = \int_0^\infty t^2 dF(t)/(2/\lambda).$$

We are interested in the joint availability ${}_kA_{joint}(x,t) = P\{I_k(t) = 1, I_k(t+x) = 1\}$ $(k = 0, 1)$ at times t and $t+x$, i.e., the joint availability is the probability that the system is up at t and again up at $t+x$ starting from state k at time 0. From renewal theoretic argument, we have the following conditional probability

(4.2.22) $$P\{I_k(t+x) = 1 | I_k(t) = 1\}$$
$$= \frac{R_k(x,t)}{A_k(t)} + \int_0^t A_0(x-u)dP\{\gamma_k(t) \le u\},$$

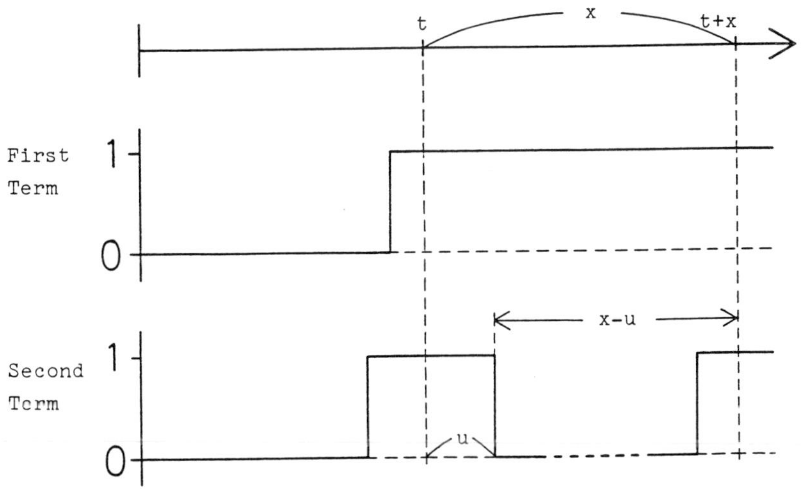

Fig. 4.2.1. Sample functions of $P\{I_k(t+x) = 1 | I_k(t) = 1\}$.

where the first term of the right-hand side denotes the probability that the system is up during $[t, t+x]$ and the second term the probability that the system is up at time t, becomes down less than or equal to $u+x$, and follows the behavior of $A_0(t-u)$ during the remaining time $(t+u, t+x]$ (see Fig. 4.2.1). The joint availability is given by

(4.2.23) $$ {}_1A_{joint}(x,t) $$

$$ = P\{I_k(t+x) = 1,\ I_k(t) = 1\} $$

$$ = P\{I_k(t+x) = 1 \mid I_k(t) = 1\}P\{I_k(t) = 1\} $$

$$ = R_1(x,t) + A_1(t)\int_0^x A_0(x-u)dP\{\gamma_1(x) \le u\} $$

(4.2.24) $$ {}_0A_{joint}(x,t) $$

$$ = R_0(x,t) + A_0(t)\int_0^t A_0(x-u)dP\{\gamma_0(t) \le u\}. $$

It is easy to show that

(4.2.25) $$ \lim_{x\to\infty} {}_kA_{joint}(x,t) = \frac{\lambda}{\lambda + \mu}A_k(t) $$

and

(4.2.26) $$ \lim_{t\to\infty}\lim_{x\to\infty} {}_kA_{joint}(x,t) = \left(\frac{\mu}{\lambda + \mu}\right)^2 . $$

We now give the limiting joint availability:

$$(4.2.27) \qquad A_{joint}(x) = \lim_{t\to\infty} {}_kA_{joint}(x,t)$$

$$= R(x) + \frac{\mu}{\lambda + \mu}\int_0^x [\bar{F}(u)/(1/\lambda)]A_0(x-u)du,$$

which can be easily derived from renewal theoretic arguments and the preceding results. Noting $\Psi(x) = \int_0^x [\bar{F}(u)/(1/\lambda)]du$, we have

$$(4.2.28) \qquad A_{joint}(x) = \frac{\mu}{\lambda + \mu}[1 - \bar{A}_0 * \Psi(x)]$$

$$= \frac{\mu}{\lambda + \mu}[1 - \Psi(x) + \Psi * A_0(x)] ,$$

where the bracket denotes the pointwise availability if the first up time distribution is $\Psi(t)$, and $\mu/(\lambda + \mu)$ is the limiting availability. That is, (4.2.28) denotes the probability that the system is up at some point (i.e., in the steady-state) and after that the system is up at time x with the initial up time distribution $\Psi(t)$.

The autocovariance of the indicator variable $I_k(t)$ is given by

$$(4.2.29) \qquad \mathrm{Cov}\{I_k(t),\ I_k(t+x)\} = A_k(x,t) - A_k(t)A_k(t+x) \qquad (k = 0,\ 1),$$

and the limiting autocovariance is given by

$$(4.2.30) \qquad \lim_{t\to\infty} \mathrm{Cov}\{I_k(t),\ I_k(t+x)\} = A_k(x) - \left(\frac{\mu}{\lambda + \mu}\right)^2 .$$

Note that the limiting variance of $I_k(t)$ is

$$(4.2.31) \qquad \lim_{t\to\infty} \mathrm{Var}\{I_k(t)\} = \frac{\lambda\mu}{(\lambda + \mu)^2} .$$

Thus the limiting autocorrelation of the indicator variable is given by

$$(4.2.32) \qquad \rho(x) = \frac{A(x) - [\mu/(\lambda + \mu)]^2}{\lambda\mu/(\lambda + \mu)^2} .$$

The corresponding spectral density is

$$(4.2.33) \qquad \frac{1}{2\pi}\int_{-\infty}^{\infty} e^{-i\omega x}\rho(x)dx ,$$

which can be applied for spectral analysis.

We finally note that if we assume the stationary alternating renewal process (see Sections 3.3 and 3.4), the limiting results are obtained for any time t.

4.3 Replacement Models

In this section we study several basic replacement models. As shown in Section 4.1, we have introduced two maintenances: Corrective maintenance and preventive maintenance. We are very much interested in the latter maintenance by which the availability is increased and/or the expected cost is reduced.

Throughout this section we assume that the lifetime is governed by the known distribution $F(t)$ $(t \geq 0)$ with finite mean $1/\lambda$. We are interested in minimizing the expected cost rate (i.e., the expected cost per unit time in the steady-state) or maximizing the limiting availability.

4.3.1 Age Replacement Models

Consider a one-unit system whose distribution is given by $F(t)$ $(t \geq 0)$ with finite mean $1/\lambda$. We assume that there are an infinite number of identical units available for replacement. Consider a renewal process with interarrival time distribution $F(t)$ which corresponds to the process of a one-unit system with corrective maintenance. We define two scheduled maintenance policies (see Barlow and Proschan (1975), p.158).

(4.3.1) Definition Under an <u>age replacement</u> policy, a unit is replaced upon failure or at age t_0, whichever comes first.

(4.3.2) Definition Under a <u>block replacement</u> policy, the unit in operation is replaced upon failure and at times T, $2T$, $3T$, ...

In this subsection we are concerned with age replacement models and in the next subsection we will discuss block replacement models.

Under an age replacement policy, a unit is replaced upon failure or at age t_0, whichever comes first. Such a renewal process becomes a truncated renewal process with interarrival times $\{X_k, t_0\}$ $(k = 1, 2, \ldots)$. The stochastic behavior of such a process can bo analyzed by using the results of renewal processes (see the details by Barlow and Proschan (1965, 1975)).

We are just interested in the expected cost per unit time in the steady-state which is referred to as the <u>expected cost rate</u> in the sequel. Let c_1 and c_2 denote the costs of failure (corrective maintenance) and scheduled replacement (preventive maintenance), respectively. It is natural to assume that

(4.3.3) $$c_1 > c_2.$$

The expected cost rate can be given by the expected cost per cycle divived by the expected duration per cycle (see Theorem (2.3.96) and Example (2.3.99)). The expected cost rate is given by

(4.3.4) $$C(t_0) = [c_1 P\{X_k \le t_0\} + c_2 P\{X_k > t_0\}]/E[\min\{X_k, t_0\}] ,$$

since a cycle terminates at $\min\{X_k, t_0\}$ $(k = 1, 2, \ldots)$ and repeats itself again and again. Noting that

(4.3.5) $$E[\min\{X_k, t_0\}]$$
$$= \int_0^{t_0} t dF(t_0) + t_0 \bar{F}(t_0)$$
$$= \int_0^{t_0} \bar{F}(t) dt,$$

we have

(4.3.6) $$C(t_0) = [c_1 F(t_0) + c_2 \bar{F}(t_0)] / \int_0^{t_0} \bar{F}(t) dt,$$

where $\bar{F}(t) \equiv 1 - F(t)$, in general. Assuming that the density $f(t)$ of $F(t)$ exists, differentiating $C(t_0)$ and equating it zero, we have

$$(4.3.7) \qquad r(t_0)\int_0^{t_0} \bar{F}(t)dt - F(t_0) = c_2/(c_1 - c_2),$$

where $r(t) = f(t)/\bar{F}(t)$, the failure rate. Equation (4.3.7) is a nonlinear equation with respect to $t_0 \geq 0$. Differentiating the right-hand side in (4.3.7), we have

$$(4.3.8) \qquad r'(t_0)\int_0^{t_0} \bar{F}(t)dt$$

which is non-negative (i.e., the right-hand side of (4.3.7) is a monotone increasing function) if we assume $r'(t_0) \geq 0$, i.e., $F(t_0)$ is IFR. Noting that the right-hand side in (4.3.7) is increasing with the right-hand side = 0 at t_0 = 0 and the right-hand side tending to $r(\infty)/\lambda - 1$ as $t_0 \to \infty$ if $F(t)$ is IFR, we have the following theorem:

(4.3.9) Theorem (i) If $r(\infty) = \lim_{t\to\infty} r(t)$ exists and $r(\infty) > K$, there exists a finite t_0 such that $C(\infty) > C(t_0)$, where $K = \lambda c_1/(c_1 - c_2)$.

(ii) If $r(t)$ is continuous and monotonely increasing with $r(\infty) > K$, there exists a finite and unique t_0 such that

$$(4.3.10) \qquad r(t_0)\int_0^{t_0} \bar{F}(t)dt - F(t_0) = c_2/(c_1 - c_2),$$

and

$$(4.3.11) \qquad C(t_0) = (c_1 - c_2)r(t_0^*).$$

(iii) If $r(t)$ is continuous and monotonely increasing with $r(\infty) > K$, there exists a finite and unique $\bar{t}_0$ such that $r(\bar{t}_0) = K$, where $\bar{t}_0 > t_0^*$, and $\bar{t}_0$ is an upper bound of the optimal t_0^*.

(4.3.12) Example The lifetime distribution is assumed to be a gamma distribution of order 2: $F(t) = 1 - (1 + 2\lambda t)e^{-2\lambda t}$, where $1/\lambda$ is the mean lifetime. The failure rate is given by $r(t) = 4\lambda^2 t/(1 + 2\lambda t)$ and $r(\infty) = 2\lambda$. If $r(\infty) = 2\lambda > K = \lambda c_1/(c_1 - c_2)$, i.e., $c_1/2 > c_2$, then there exists a finite and unique t_0^* satisfying equation (4.3.10), and $C(t_0^*) = 4\lambda^2 t_0^*(c_1 - c_2)/(1 + 2\lambda t_0^*)$. Otherwise, if $c_1/2 \le c_2$, the optimal $t_0^* = \infty$, i.e., no scheduled maintenance (only corrective maintenance).

Let us next consider a discounted version of the age replacement model. Let α be the <u>discount</u> <u>rate</u> of exponential type. That is, a unit cost is discounted to $e^{-\alpha t}$ after time t. For the discounted model we introduce the total discounted expected cost for an infinite time span. Then we have the following recursive equation:

(4.3.13) $$C(\alpha;t_0) = c_1\int_0^{t_0} e^{-\alpha t}dF(t) + c_2 e^{-\alpha t_0}\bar{F}(t_0) + [\int_0^{t_0} e^{-\alpha t}dF(t) + e^{-\alpha t_0}\bar{F}(t_0)]C(\alpha;t_0),$$

which is just similar to renewal type equation discussed in Section 2.3. Note that

(4.3.14) $$1 - [\int_0^{t_0} e^{-\alpha t}dF(t) + e^{-\alpha t_0}\bar{F}(t_0)] = \alpha\int_0^{t_0} e^{-\alpha t}\bar{F}(t)dt$$

we have

(4.3.15) $$C(\alpha;t_0) = \frac{c_1\int_0^{\infty} e^{-\alpha t}dF(t) + c_2 e^{-\alpha t_0}\bar{F}(t_0)}{\alpha\int_0^{\infty} e^{-\alpha t}\bar{F}(t)dt} .$$

Let

$$(4.3.16) \qquad K(\alpha) = \frac{c_1F^*(\alpha) + c_2[1 - F^*(\alpha)]}{(c_1 - c_2)[1 - F^*(\alpha)]/\alpha}$$

which corresponds to K in Theorem (4.3.4) for the discounted version, where $F^*(\alpha) = \int_0^\infty e^{-\alpha t}dF(t)$, the Laplace-Stieltjes transform evaluated at $s = \alpha$. Let

$$(4.3.17) \quad r(t_0)\int_0^{t_0} e^{-\alpha t}\bar{F}(t)dt - \int_0^{t_0} e^{-\alpha t}dF(t) = c_2/(c_1 - c_2),$$

which corresponds to equation (4.3.10) for the nondiscounted version. We can easily verify the similar theorem of Theorem (4.3.9) for the discounted model. In particular, if there exists a finite and unique t_0^*, we have

$$(4.3.18) \qquad C(\alpha;t_0^*) = (c_1 - c_2)r(t_0^*)/\alpha - c_2.$$

We note that the discounted model includes the nondiscounted one by perturbation such as $\lim_{\alpha\to 0} \alpha C(\alpha;t_0) = C(t_0)$ and $\lim_{\alpha\to 0} K(\alpha) = K$.

The above two models are based on the continuous time random variable with distribution F(t). We next consider the discrete version of the age replacement model. Let p(n) (n = 1, 2, ...) be the probability mass function of the lifetime of a unit with finite mean

$$(4.3.19) \qquad 1/\lambda = \sum_{n=1}^{\infty} np(n).$$

The expected cost rate is given by

$$(4.3.20) \qquad C(n_0) = \frac{c_1 \sum_{j=1}^{n_0} p(j) + c_2 \sum_{j=n_0+1}^{\infty} p(j)}{\sum_{i=1}^{n_0} \sum_{j=i}^{\infty} p(j)}$$

which is a function of the replacement interval $n_0 \geq 1$. Noting that the failure rate $r(n)$ is defined by

$$(4.3.21) \qquad r(n) = p(n)/\sum_{j=n}^{\infty} p(j) \qquad (n = 0, 1, 2, \ldots)$$

(see Equation (1.4.37)), we have the following theorem for the discrete version:

(4.3.22) Theorem (i) If $r(\infty) = \lim_{n\to\infty} r(n)$ exists and $r(\infty) > K$, there exists a finite n_0 such that $C(\infty) > C(n_0)$, where $K = \lambda c_1/(c_1 - c_2)$.

(ii) If $r(n)$ is monotonely increasing with $r(\infty) > K$, there exists a finite and unique n_0 $(1 \leq n_0 < \infty)$ such that $L(n_0) \geq c_2/(c_1 - c_2)$ and $L(n_0 - 1) < c_2/(c_1 - c_2)$ where

$$(4.3.23) \qquad L(n_0) = r(n_0+1) \sum_{i=1}^{n_0} \sum_{j=i}^{\infty} p(j) - \sum_{j=1}^{n_0} p(j) \quad (n_0 = 1, 2, \ldots)$$

and

$$(4.3.24) \qquad (c_1 - c_2) r(n_0^*) < C(n_0^*) < (c_1 - c_2) r(n_0^*+1).$$

(iii) If $r(n)$ is monotonely increasing with $r(\infty) > K$, there exists a finite and unique $\bar{n}_0$ such that $\bar{n}_0$ is the minimum solution satisfying $r(n_0+1) \geq K$, where $\bar{n}_0 > n_0^*$ and $\bar{n}_0$ is an upper bound of the optimal n_0^*.

Let us finally consider the discounted version of the

discrete time age replacement model. Let β be the discount factor for each unit time. That is, a unit cost is discounted β^n after n unit times. Just similar to the above two models, we have the following total discounted cost:

$$(4.3.25) \quad C(\beta;n_0) = \frac{c_1 \sum_{j=1}^{n_0} \beta^j p(j) + c_2 \beta^N \sum_{j=n_0+1}^{\infty} p(j)}{[(1-\beta)/\beta] \sum_{i=1}^{n_0} \beta^i \sum_{j=i}^{\infty} p(j)} .$$

We can similarly obtain the corresponding theorem for the discounted version of the discrete time age replacement model. It is noted from perturbation that

$$(4.3.26) \qquad C(n_0) = \lim_{\beta\to 1} (1 - \beta) C(\beta;n_0).$$

4.3.2 Block Replacement Models

For the age replacement model we have to observe the age of a unit, which is sometimes difficult to administrate. However, for the block replacement model we don't need to observe the age of a unit, but replace at T, 2T, 3T, ..., which is easier to administrate in general (see Definitions (4.3.1) and (4.3.2)). In this subsection we discuss the following three variations of the block replacement models:

(I) A failed unit is replaced instantaneously at failure.
(II) A failed unit remains failure until the next

scheduled replacement.

(III) A failed unit undergoes a minimal repair.

We call Models I, II and III for the above model, respectively. In general, a typical block replacement model is Model I which is well-known so far. Model III was discussed by Barlow and Hunter (1960) and called periodic replacement with minimal repair at failure, where we assume that after each failure only minimal repair is made so that system failure rate $r(t)$ is not disturbed. Model II was discussed by Nakagawa (1979) and is a simple variation of the block replacement models.

Model I

A failed unit is replaced by a new unit during the replacement interval T and at $T, 2T, 3T, \ldots,$ the scheduled replacement for the non-failed unit is made. Then the expected cost rate is

$$(4.3.27) \qquad C_1(T) = [c_1 M(t) + c_2]/T$$

$$= [c_1 \int_0^T m(t)dt + c_2]/T,$$

where

$$(4.3.28) \qquad M(t) = \sum_{n=1}^{\infty} F^{(n)}(t)$$

is the renewal function with underlying distribution $F(t)$, $m(t) = dM(t)/dt$ is the renewal density, c_1 is the cost of replacement for the failed unit, and c_2 is the cost of scheduled replacement for the non-failed unit. To minimize the expected cost rate in (4.3.27), we have the following equation by differentiating $C(T)$ and equating it zero:

$$(4.3.29) \qquad Tm(T) - \int_0^T m(t)dt = c_2/c_1.$$

This is a necessary condition that there exist a finite T^*, and, in this case, the resulting expected cost rate is

$$(4.3.30) \qquad C_1(T^*) = c_1 m(T^*).$$

In general, specifying the underlying distribution $F(t)$, we can obtain the renewal function $M(t)$ and renewal density $m(t)$, by which we can solvo the nonlinear equation (4.3.29).

(4.3.31) Example We assume that the lifetime distribution is a gamma distribution of order 2 (see Example (2.3.12)):

$$(4.3.32) \qquad \bar{F}(t) = (1 + at)e^{-at},$$

$$(4.3.33) \qquad m(t) = (a/2)\,(1 - e^{-2at}).$$

Note that the mean lifetime is $1/\lambda = 2/a$. Equation (4.3.30) becomes

$$(4.3.34) \quad (1/4)[1 - e^{-2aT} - 2aTe^{-2aT}] = c_2/c_1.$$

Thus, if $c_1/c_2 \geq 1/4$, we should make no scheduled replacement, i.e., a unit is replaced only at failure. If $c_1/c_2 < 1/4$, there exists a finite and unique T^* which minimizes (4.3.27) and the resulting expected cost rate is

$$(4.3.35) \qquad C_1(T^*) = (c_1 a/2)\,[1 - e^{-2aT^*}].$$

Model II

For the first model we have assumed that a failed unit is detected instantaneously and its replacement is made instantaneously just after detection. For Model II, we assume that failure is detected only at $T, 2T, 3T, \ldots$.

The unit is always replaced at $T, 2T, 3T, \ldots$, but is not replaced at failure, and the unit remains failure for the time duration from the occurrence of failure to its detection. The expected duration from the occurrence of failure to its detection per cycle is given by

(4.3.36) $$\int_0^T (T - t)dF(t) = \int_0^T F(t)dt.$$

Thus, the expected cost rate is

(4.3.37) $$C_2(T) = [c_1 \int_0^T F(t)dt + c_2]/T,$$

where c_1 is the cost of failure per unit time and c_2 is the cost of replacement for the non-failed unit. Comparing (4.3.37) to (4.3.27), we have the similar expected cost rate except the integrand. The similar discussions developed in Model I will be done.

(4.3.38) Example Noting that

(4.3.39) $$\lim_{T\to\infty} \int_0^T [F(T) - F(t)]dt = \int_0^\infty \bar{F}(t)dt,$$

we can show that if $1/\lambda > c_2/c_1$, there exists an optimal

T^* which is a unique and finite solution to

$$(4.3.40) \qquad \int_0^T [F(T) - F(t)]dt = c_2/c_1.$$

If we assume the same distribution $F(t)$ in (4.3.32), and $\int_0^\infty \bar{F}(t)dt = 2/a \leq c_2/c_1$, we should make no scheduled replacement. If $2/a > c_2/c_1$, there exists a unique T^* satisfying

$$(4.3.41) \quad (1/2)[2 - e^{-aT}(2 + 2aT + a^2T^2)] = c_2/c_1 ,$$

and the resultant expected cost rate is

$$(4.3.42) \qquad C_2(T^*) = c_1[1 - (1 + aT^*)e^{-aT^*}].$$

Model III

We assume that a minimal repair is made when the unit fails and the failure rate is not disturbed by each repair. If we consider a stochastic process $\{N(t), t \geq 0\}$ in which $N(t)$ represents the number of minimal failures up to time t, the process $\{N(t), t \geq 0\}$ is governed by a nonhomogeneous Poisson process with mean value function

$$(4.3.43) \qquad R(t) = \int_0^t r(x)dx,$$

which is the hazard function (see Subsection 1.4.1). Noting this fact, we have the following expected cost rate for Model III:

(4.3.44) $$C_3(T) = [c_1 \int_0^T r(x)dx + c_2]/T,$$

where c_1 and c_2 are the costs of minimal repair for the failed unit and of replacement for the non-failed unit, respectively. Comparing (4.3.44) to (4.3.27), we have also the similar expected cost rate except the integrand. The similar discussions developed in Models I and II will be done.

(4.3.45) Example If we assume the same distribution $F(t)$ in (4.3.32), we have

(4.3.46) $$r(t) = a^2t/(1 + at).$$

There exists a unique T^* satisfying

(4.3.47) $$\log (1+aT) - aT/(1+aT) = c_2/c_1,$$

and the resultant expected cost rate is

(4.3.48) $$C_3(T^*) = c_1a^2T^*/(1 + aT^*).$$

We summarize three block replacement models: Models I, II and III. The expected cost rate is given by

(4.3.49) $$C_i(T) = [c_1 \int_0^T \phi_i(t)dt + c_2]/T,$$

where

(4.3.50) $$\phi_1(t) = m(t),$$

the renewal density for Model I,

(4.3.51) $\phi_2(t) = F(t)$,

the lifetime distribution for Model II, and

(4.3.52) $\phi_3(t) = r(t)$,

the failure rate for Model III. We also note the dimension of cost c_1 for each model: The cost c_1 for Models I and II is incurred for each failure, but the cost c_1 for Model II is incurred for each unit time when the unit remains failure from the occurrence of failure to its detection.

To minimize $C_i(T)$ $(i - 1, 2, 3)$, we should solve the following equation

(4.3.53) $\int_0^T [\phi_i(T) - \phi_i(t)]dt = c_2/c_1$,

whose solution T^* satisfies a necessary condition of the minimum. If the solution T^* to (4.3.53) also satisfies a sufficient condition, the optimal expected cost rate is given by

(4.3.54) $C_i(T^*) = c_1\phi_i(T^*)$.

In practice, if we can specify $\phi_i(t)$, we can analytically show the necessary and sufficient condition if it exists for most cases.

We next consider the discounted version of the above block replacement models. Let α be the discount rate of exponential type and $C_i(\alpha;T)$ be the total expected cost for an infinite time span for Model i $(i = 1, 2, 3)$, respectively. Then we have

$$(4.3.55) \qquad C_i(\alpha;T) = [c_1 \int_0^T e^{-\alpha t} \phi_i(t) + c_2]/(1 - e^{-\alpha T}).$$

We also have the following equation by differentiating $C_i(\alpha;T)$ and equating it zero:

$$(4.3.56) \qquad \int_0^T e^{-\alpha t}[\phi_i(T) - \phi_i(t)]dt = c_2/c_1 ,$$

which is a necessary condition of the minimum that there exists a finite T^* and, in this case, the resulting expected cost rate is

$$(4.3.57) \qquad C_i(\alpha;T^*) = (c_1/\alpha)\, \phi_i(T^*) - c_2.$$

Note that, from perturbation theory, $\lim_{\alpha \to 0} \alpha C_i(\alpha;T) = C_i(T)$, which is the expected cost rate for the nondiscounted version.

Let us finally consider the discrete version for block replacement models above. The expected cost rate is given by

$$(4.3.58) \qquad C_i(N) = [N \sum_{j=1}^{N} \phi_i(j) + c_2]/N,$$

where

$$(4.3.59) \qquad \phi_1(j) = m(j)$$

the discrete renewal density for Model I,

$$(4.3.60) \qquad \phi_2(j) = \sum_{i=1}^{j} p(i)$$

the distribution for Model II, and

$$(4.3.61) \qquad \phi_3(j) = r(j) = p(j)/ \sum_{i=j}^{\infty} p(i)$$

the discrete failure rate for Model III. The inequalities $C_i(N+1) \geq C_i(N)$ and $C_i(N) > C_i(N-1)$ give

$$L_i(N) \geq c_2/c_1 \quad \text{and} \quad L_i(N-1) < c_2/c_1 \tag{4.3.62}$$

where

$$L_i(N) = \begin{cases} N\phi_i(N+1) - \sum_{j=1}^{N} \phi(j) & (N = 1, 2, \ldots) \\ 0 & (N = 0). \end{cases} \tag{4.3.63}$$

The resulting expected cost rate is

$$c_1\phi_i(N^*) < C_i(N^*) \leq c_1\phi_i(N^*+1) \tag{4.3.64}$$

for Models I, II and III, respectively.

4.4 Ordering Models

In an age replacement model, there are an unlimited number of units available immediately for replacement. However, it is not always true and we assume here that a unit for each replacement can be supplied by order with lead time. That is, only by order we can obtain a unit for each replacement after a lead time. The policy in which we are interested is when we should order a unit for each replacement. Thus, we call such a policy an ordering policy with lead time. We call such a model an ordering model in general.

Let us introduce the notations and the assumptions used here. Assume that we can obtain an identical unit for each replacement by order. The random variable X denotes the lifetime of an operating unit with arbitrary distribution $F(t)$ $(t \geq 0)$ having finite mean $1/\lambda$. It is assumed that the failure of an operating unit can be immediately detected. Let t_0 $(t_0 \geq 0)$ be the ordering time of a unit, measured from the installation of an operating unit. Let us next introduce the cost structures. It incurs a cost c_1 for each expedited order after the failure of an operating unit. It also incurs a cost c_2 for each regular order before the failure of an operating unit. It is plausible to assume $c_1 > c_2$, since an expedited order is much more expensive than a regular order. The ordered unit can be obtained after a constant lead time $L \geq 0$, and takes over its operation immediately or after in inventory up to the failure of an operating unit. If an operating unit fails when an ordered unit is not delivered, then it cannot operate until the ordered unit is delivered. This incurs a cost k_1 per unit time for shortage. On the other hand, if an ordered unit is delivered before it is needed, then it is put into inventory. This incurs a constant cost k_2 per unit time for inventory.

Two ordering policies can be considered depending upon the situation of inventory of an ordered unit: For Model I, an ordered unit is put into inventory up to the failure of an operating unit if an operating unit is alive, and, of course, an ordered unit takes over its operation immediately just after delivery if an operating unit has been failed so far. For Model II, an ordered unit takes over its operation just after delivery, irrespective of any situation of an operating unit.

4.4.1 Model I

We consider an infinite planning horizon. For an infinite planning horizon, it is appropriate to adopt an expected cost per unit time in the steady-state, i.e., the expected cost rate as an objective function.

Consider one cycle from the beginning of the operating unit to its replacement. Then, the expected cost of one cycle is given by the sum of tho following three costs: (i) The expected shortage cost is

$$(4.4.1) \qquad k_1\left[\int_0^{t_0} L\,dF(t) + \int_{t_0}^{t_0+L} (t_0 + L - t)\,dF(t)\right] = k_1 \int_{t_0}^{t_0+L} F(t)\,dt,$$

since the shortage cost is proportional to the shortage time. (ii) The expected inventory cost is

$$(4.4.2) \qquad k_2 \int_{t_0+L}^{\infty}(t - t_0 - L)\,dF(t) = k_2 \int_{t_0+L}^{\infty}\bar{F}(t)\,dt ,$$

where $\bar{F}(t) \equiv 1 - F(t)$. (iii) The expected ordering cost is

$$(4.4.3) \qquad c_1F(t_0) + c_2\bar{F}(t_0).$$

Further, the mean time of one cycle is

$$(4.4.4)\quad \int_0^{t_0} (L + t)dF(t) + \int_{t_0}^{t_0+L} (t_0 + L)dF(t) + \int_{t_0+L}^{\infty} t\,dF(t).$$

$$= 1/\lambda + \int_{t_0}^{t_0+L} F(t)dt.$$

Thus, the total expected cost per one cycle is

$$(4.4.5)\quad C_1(t_0) = \frac{[k_1\int_{t_0}^{t_0+L} F(t) + k_2\int_{t_0+L}^{\infty} \bar{F}(t)dt + c_1F(t_0) + c_2\bar{F}(t_0)]}{[1/\lambda + \int_{t_0}^{t_0+L} F(t)dt]}.$$

The expected cost and the mean time at each cycle are the same, and hence, $C_1(t_0)$ is equal to the expected cost rate (see Theorem (2.3.96)).

Of our interest is to obtain the optimum ordering time t_0^* minimizing the expected cost $C_1(t_0)$ in (4.4.5) under the assumption that $c_1 > c_2$. It is assumed that there exists the density $f(t)$ of the failure time distribution $F(t)$. Let $F(L|t_0) \equiv [F(t_0+L)-F(t_0)]/\bar{F}(t_0)$ and $r(t_0)$ $f(t_0)/\bar{F}(t_0)$. Then, both $F(L|t_0)$ and $r(t_0)$ are called by the failure rates and they have the same properties of the failure rates, i.e., $F(L|t_0)$ is increasing (decreasing) if and only if $r(r_0)$ is increasing (decreasing), respectively (see Barlow and Proschan (1965), p.23). We restrict ourselves to the case in which the failure rate is continuous and has a monotone property. Further, we assume that $C_1(\infty) < k_1$, i.e., $c_1 < k_1/\lambda$, because the expected cost of the system in which the order is made after failure of the original unit would be less than that of the system which remains inoperative forever. Let

$$(4.4.6)\quad q_1(t)$$

$$= F(L|t)[k_1/\lambda + k_2(L + \int_0^t \bar{F}(u)du) - c_1F(t) - c_2\bar{F}(t)]$$

$$+ [(c_1 - c_2)r(t) - k_2][1/\lambda + \int_t^{t+L} F(u)du],$$

for simplicity of equations. Then, we have the following theorem:

(4.4.7) Theorem Assume that $c_1 < k_1/\lambda$: (1) Suppose that $r(t)$ is monotonely increasing. If $q_1(0) < 0$ and $q_1(\infty) > 0$, then t_0^* exists uniquely on $(0, \infty)$ as the solution to $q_1(t_0) = 0$. Otherwise, $t_0^* = \infty$ or 0 according as $q_1(\infty) \leq 0$ or $q_1(0) \geq 0$, respectively.
(2) Suppose that $r(t)$ in non-increasing. Then, $t_0^* = \infty$ (0) if

$$(4.4.8)\qquad (1/\lambda + L)(k_2\int_L^\infty \bar{F}(t)dt - c_1 + c_2)$$

$$\geq (<)\ (k_1/\lambda - c_1)\int_0^L \bar{F}(t)dt.$$

Proof. Differentiating $C_1(t_0)$ with respect to t_0 and setting it equal to zero, we have $q_1(t_0) = 0$. Further, from the assumption that $c_1 < k_1/\lambda$, $q_1(t_0)$ is monotonely increasing (non-increasing) if $r(t_0)$ is monotonely increasing (non-increasing), respectively.

First suppose that $r(t_0)$ is monotonely increasing. If $q_1(0) < 0$ and $q_1(\infty) > 0$, then from the monotonicity and the continuity of $q_1(t_0)$, t_0^* exists uniquely on $(0, \infty)$ as the solution to $q_1(t_0) = 0$, which minimizes the expected cost $C_1(t_0)$. Further, it is easily shown that if $q_1(\infty) \leq$

0 then $t_0^* = 0$.

Next, suppose that $r(t_0)$ is non-increasing. Then, $q_1(t_0)$ is also non-increasing. Thus, it is easily seen that $C_1(0)$ or $C_1(\infty)$ is not greater than $C_1(t_0)$ for any t_0. Therefore, we have $t_0^* = \infty$ if $C_1(\infty) \leq C_1(0)$, i.e.,

$$(4.4.9) \qquad (1/\lambda + L)(k_2 \int_L^\infty \bar{F}(t)dt - c_1 + c_2) \geq (k_1/\lambda - c_1) \int_0^L \bar{F}(t)dt,$$

and vice versa.

In case of $q_1(0) < 0$ and $q_1(\infty) > 0$ of (1) in Theorem (4.4.7), the expected cost is given by

$$(4.4.10) \qquad C_1(t_0^*) = k_1 + k_2 - [k_2 - r(t_0^*)(c_1 - c_2)]/F(L|t_0^*).$$

Further, the ordering policy when $t_0^* = \infty$ represents that the order of the spare unit is made immediately after failure of the original unit and the policy when $t_0^* = 0$ represents that its order is made at the same time of the beginning of the original unit.

In the above theorem, we have assumed that $c_1 < k_1/\lambda$. Of course, we can also prove that the above theorem under the weaker condition than $c_1 < k_1/\lambda$, for instance, $c_1 < k_1/\lambda + k_2L$. However, in actual situations, any ordering policy might be better than no order, in which the system has been remaining inoperative forever. It would be waste to discuss an optimum policy under the assumption that $c_1 \geq k_1/\lambda$.

In case of $q_1(0) < 0$ and $q_1(\infty) > 0$ of (1) in

Theorem (4.4.7), we can obtain the following upper limit of the optimum ordering time t_0^*. This could be applied to compute t_0^* by the successive approximations (see numerical examples below).

(4.4.11) Theorem Suppose that $c_1 < k_1/\lambda$, $q_1(0) < 0$, $q_1(\infty) > 0$ and $r(t)$ is monotonely increasing. If $\bar{t}_0$ is a solution to $h_1(t_0) = 0$ then $\bar{t}_0$ exists uniquely (possibly infinite) and $t_0^* < \bar{t}_0$, where

$$(4.4.12) \qquad h_1(t_0) = F(L|t_0)[k_1/\lambda + k_2L - c_2]$$

$$+ [(c_1 - c_2)r(t_0) - k_2][1/\lambda + \int_0^L F(t)dt].$$

Proof. We can easily obtain

$$(4.4.13) \qquad r(t_0) > F(t_0)/\int_0^{t_0} \bar{F}(t)dt,$$

$$(4.4.14) \qquad F(L|t_0) > [\int_{t_0}^{t_0+L} F(t)dt - \int_0^L F(l)dl]/\int_0^{t_0} \bar{F}(t)dt,$$

since the failure rate is monotonely increasing. Thus, we have $q_1(t_0) > h_1(t_0)$ for $0 < t_0 < \infty$. If $\bar{t}_0$ is a solution to $h_1(t_0) = 0$ then $\bar{t}_0$ is unique because $h_1(t_0)$ is monotonely increasing and $t_0^* < \bar{t}_0$.

In the remaining of this subsection we discuss the discounted and discrete versions of an ordering model. Let us first consider the discounted model. Let α be the discount rate of exponential type. Then the total discounted expected cost $C_1(\alpha;t_0)$ is given by

$$(4.4.15) \quad C_1(\alpha\,; t_0)$$

$$= \{c_1 \int_0^{t_0} e^{-\alpha(t+L)} dF(t) + c_2 e^{-\alpha(t_0+L)} \bar{F}(t_0)$$

$$+ k_1[(1 - e^{-\alpha L}) \int_0^{t_0} e^{-\alpha t} F(t)dt + \int_{t_0}^{t_0+L} e^{-\alpha t} F(t)dt]$$

$$+ k_2 \int_{t_0+L}^{\infty} e^{-\alpha t} \bar{F}(t)dt\}/\{1 - e^{-\alpha L}$$

$$+ \alpha[\int_0^{t_0} e^{-\alpha(t+L)} \bar{F}(t)dt + \int_{t_0+L}^{\infty} e^{-\alpha t} \bar{F}(t)dt]\}$$

(see the details of derivation by Osaki (1979)). We can develop the similar theorems for the discounted model. In particular, if there exists a finite and unique t_0^* satisfying the equation under suitable assumptions, then the resulting total discounted expected cost is

$$(4.4.16) \quad C_1(\alpha\,; t_0^*)$$

$$= [(c_1 - c_2) r(t_0^*) - \alpha c_2 + (k_1 + k_2) F(L|t_0^*) - k_2] / [\alpha F(L|t_0^*)] .$$

It is clear from perturbation theory that

$$(4.4.17) \qquad C_1(t_0) = \lim_{\alpha \to 0} \alpha C_1(\alpha; t_0)$$

which is the expected cost rate for the nondiscounted model.

Let us next consider the discrete model. Let $p(j)$ ($j = 1, 2, \ldots$) be the probability mass function of the discrete lifetime distribution. Note that the failure rate is

$$(4.4.18) \qquad r(N) = p(N) / \sum_{j=N}^{\infty} p(j)$$

and the conditional probability $F(L|N)$ is

$$F(L|N) = [\sum_{j=1}^{N+L} p(j) - \sum_{j=1}^{N} p(j)] / \sum_{j=N}^{\infty} p(j), \tag{4.4.19}$$

and

$$1/\lambda = \sum_{j=1}^{\infty} jp(j). \tag{4.4.20}$$

Then we can obtain the following expected cost rate as a function of the discrete ordering time N ($N = 1, 2, \ldots$):

$$\begin{aligned} C_1(N) &= [c_1 \sum_{j=1}^{N} p(j) + c_2 \sum_{j=N+1}^{\infty} p(j) \\ &\quad + k_1 \sum_{i=N+1}^{N+L} \sum_{j=1}^{i-1} p(j) + k_2 \sum_{i=N+L+1}^{\infty} \sum_{j=i}^{\infty} p(j)] \\ &\quad /[1/\lambda + \sum_{j=N}^{N+L} \sum_{j=1}^{i-1} p(j)]. \end{aligned} \tag{4.4.21}$$

Using this expected cost rate, we can develop the similar theorems which have be done in Theorems (4.4.7) and (4.4.11).

4.4.2 Model II

In Model I, it has been assumed that the delivered unit is put into inventory if the original unit is operative. Here, the model has the same assumptions as Model I except that the original unit is always replaced as soon as the spare unit is delivered, even if it is operating. This model is appropriate in cases such that an inventory task is very difficult or there is no place to put a spare unit in inventory.

In the model, we do not need to consider the inventory cost because of the assumption. The shortage cost is equal to (4.4.1) and hence, the total expected cost per one cycle is

$$(4.4.22) \qquad C_2(t_0) = \frac{k_1\int_{t_0}^{t_0+L} F(t)dt + c_1(t_0) + c_2\bar{F}(t_0)}{L + \int_0^{t_0} \bar{F}(t)dt}.$$

Of our interest is to obtain the optimum t_0^* minimizing the expected cost $C_2(t_0)$ in (4.4.22) under the assumption that $c_1 > c_2$. Let

$$(4.4.23) \qquad q_2(t) = [r(t) + b_1F(L|t)][L + \int_0^t \bar{F}(u)du] - F(t) - b_1\int_t^{t+L} F(u)du,$$

where $b_1 = k_1/(c_1 - c_2)$ and $b_2 = c_2/(c_1 - c_2)$. Then, from the discussion similar to previous theorems, we obtain the following theorems without proving.

(4.4.24) Theorem (1) Suppose that $r(t)$ is monotonely increasing. If $q_2(0) < b_2$ and $q_2(\infty) > b_2$, then t_0^* exists uniquely on $(0, \infty)$ as the solution to $q_2(t_0) = b_2$. Otherwise, $t_0^* = \infty$ or 0 according as $q_2(\infty) \leq b_2$ or $q_2(0) \geq b_2$, respectively.

(2) Suppose that $r(t)$ is non-increasing. Then, $t_0^* = \infty$ (0) if

(4.4.25) $$(1/\lambda + L)(k_1 \int_0^L \bar{F}(t)dt - c_1) \leq (\geq) (k_1/\lambda - c_1)L.$$

(4.4.26) Theorem Suppose that $q_2(0) < b_2$, $q_2(\infty) > b_2$ and $r(t)$ is monotonely increasing. If $\bar{t}_0$ is a solution to $h_2(t_0) = 0$ then $\bar{t}_0$ exists uniquely (possibly infinite) and $t_0^* < \bar{t}_0$, where

(4.4.27) $$h_2(t_0) = r(t_0) + b_1 R(t_0) - [b_1 \int_0^t F(t)dt + b_2]/L.$$

In case of $q_2(0) < b_2$, and $q_2(\infty) > b_2$ of (1) in Theorem (4.4.24), the expected cost is

(4.4.28) $$C_2(t_0^*) = k_1 F(L|t_0^*) + (c_1 - c_2)r(t_0^*).$$

We can also discuss the discounted and discrete versions of an ordering model. For the discounted model, the total discounted expected cost $C_2(\alpha;t_0)$ is given by

(4.4.29) $$C_2(\alpha;t_0^*)$$
$$= \{c_1 \int_0^{t_0} e^{-\alpha(t+L)} dF(t) + c_2 e^{-\alpha(t_0+L)} \bar{F}(t_0)$$
$$+ k_1[(1 - e^{-\alpha L}) \int_0^{t_0} e^{-\alpha t} F(t)dt + \int_{t_0}^{t_0+L} e^{-\alpha t} F(t)dt]\}$$
$$/[1 - e^{-\alpha(t_0+L)} - \alpha \int_0^{t_0} e^{-\alpha(t+L)} F(t)dt],$$

which can be similarly obtained as shown in (4.4.16). We can also develop the similar theorem for the discounted model. In particular, if there exists a finite and unique t_0^* satisfying the equation under suitable assumptions, then the resulting total discounted expected cost is

(4.4.30) $$C_2(\alpha\,;t_0^*) = \frac{(c_1 - c_2)r(t_0^*) - \alpha c_2 + k_1 F(L|t_0^*)}{\alpha e^{-\alpha(t_0^*+L)}}$$

It is clear from perturbation theory that

(4.4.31) $$C_2(t_0) = \lim_{\alpha\to 0} \alpha C_2(\alpha;t_0).$$

Let us next consider the discrete model. Noting the notations in $r(N)$ and $F(L|N)$, we have the following expected cost rate as a function of the discrete ordering time N $(N = 1, 2, \ldots)$:

(4.4.32) $$C_2(N) = [c_1 \sum_{j=1}^{N} p(j) + c_2 \sum_{j=N+1}^{\infty} p(j) + k_1 \sum_{i=N+1}^{N+L} \sum_{j=1}^{i-1} p(j)]/[L + \sum_{i=1}^{N} \sum_{j=i}^{\infty} p(j)].$$

We can also develop the similar theorems for the discrete model.

(4.4.33) Example We assume that

(4.4.34) $$F(t) = 1 - (1 + at)e^{-at} \qquad (t \geq 0).$$

Then

(4.4.35) $$r(t) = a^2 t/(1 + at),$$

(4.4.36) $$F(L|t) = 1 - e^{-aL} - Le^{-aL}/(1 + at),$$

(4.4.37) $$1/\lambda = 2/a.$$

It is noted that the failure time distribution is a gamma

distribution of order 2, which has a monotone increasing failure rate with $r(0) = 0$ and $r(\infty) = a$.

As an example, we consider Model II. From Theorem (4.4.24), if

$$(4.4.38) \qquad (L + 2/a)[a + b_1(1 - e^{-aL})] > 1 + b_1 L + b_2$$

and

$$(4.4.39) \qquad b_1[1/a - (2/a + 2L + aL^2)e^{-aL}] < b_2,$$

then t_0^* exists as the solution to

$$(4.4.40) \qquad A(1 + at) + Be^{-aL} = D,$$

where

$$(4.4.41) \qquad A \equiv (L + 2/a)[a + b_1(1 - e^{-aL})] - (1 + b_1 L + b_2),$$

$$(4.4.42) \qquad B \equiv 1 + b_1 Le^{-aL},$$

$$(4.4.43) \qquad D \equiv (L + 2/a)(1 + b_1 Le^{-aL}).$$

In this case, the expected cost is

$$(4.4.44) \qquad C_2(t_0^*) = k_1(1 - e^{-aL}) + [(c_1 - c_2)a^2 t_0^* - k_1 Le^{-aL}]/(1 + at_0^*).$$

Moreover, from the inequality that $t_0^* < \bar{t}_0$ in Theorem (4.4.26), we have

$$(4.4.45) \quad at_0^* < \frac{b_2 - b_1[2/a - (2/a + 2L + aL^2)e^{-aL}]}{(L + 2/a)[a + b_1(1 - e^{-aL})] - 2 - b_1L(1 + e^{-aL}) - b_2}$$

if the right-hand side is positive.

4.5 Inspection Policies

We have discussed several replacement models in which failure can be detected instantaneously except Model II of block replacement. In this section we are interested in inspection policies by Barlow and Proschan (1965).

If there is a system whose failure can be detected only by inspection, it is necessary and important to obtain the effective procedure for detecting the system failure, i.e., the so-called optimal inspection policy. If we execute the inspections to the system too frequently, we must waste the cost for inspection too much. Conversely, if we reduce the inspections, the delay time elapsed between system failure and its detection takes place, and, consequently, the cost for system failure increases. Thus, we must obtain the optimal inspection policy minimizing the total expected cost which balances the inspection and the system failure costs. From this viewpoint, many inspection policies have been discussed.

We show the following assumptions for a basic

inspection model.

1. A one-unit system is considered.
2. The system begins operation at time 0, and the planning horizon is infinite.
3. The lifetime for the system obeys an arbitrary distribution $F(t)$ with density $f(t)$.
4. The system failure is revealed only by inspection made at inspection time x_k $(k = 1, 2, 3, \ldots)$.
5. Each inspection is perfect, instantaneous, and does not cause the deterioration or failure of the system.
6. The inspection policy terminates with the detection of system failure.
7. Each inspection incurs a fixed cost c_1.
8. The time elapsed between system failure and its detection at the next inspection costs c_2 per unit time.

If we assume that $N(t)$ denotes the number of inspections during $[0, t]$ and γ_t denotes the interval between failure and its detection if failure takes place at time t, the resultant cost is $c_1[N(t)+1]+c_2\gamma_t$, where an additional inspection always follows failure. The total expected cost C to failure detection is given by

$$(4.5.1) \qquad C = \int_0^\infty \{c_1(E[N(t)]+1)+c_2E[\gamma_t]\}dF(t).$$

Let $\{x_1, x_2, x_3, \ldots\}$ $(x_1 < x_2 < x_3 < \ldots)$ be a time sequence of inspections which minimize the total expected cost C. Then we have

$$(4.5.2) \qquad C = \sum_{k=0}^{\infty} \int_{x_k}^{x_{k+1}} [c_1(k+1)+c_2(x_{k+1}-t)]dF(t)$$

where we assume $x_0 = 0$.

A necessary condition that a sequence $\{x_k\}$ is a minimum cost inspection procedure is obtained by partial differentiation $\partial C/\partial x_k = 0$ for all k:

$$(4.5.3) \qquad x_{k+1} - x_k = \frac{F(x_k) - F(x_{k+1})}{f(x_k)} - \frac{c_1}{c_2} \quad (k = 1,2,\ldots).$$

The sequence $\{x_k\}$ can be determined recursively once we specify x_1.

Barlow and Proschan (1965) proved a series of theorems to solve equation (4.5.3). In particular, they proved the following: If the lifetime distribution is IFR and $f(x) > 0$ for $x > 0$, the optimal inspection intervals are non-increasing.

We cite an example by Barlow and Proschan (1965), pp.113-114.

(4.5.4) Example Assume that the lifetime is uniformly distributed over the interval $[0, T]$, i.e., $F(t) = t/T$ and $f(t) = 1/T$ $(0 \le t \le T)$. Then from (4.5.3), we have

$$(4.5.5) \qquad x_{k+1} - x_k = c_1/c_2 = x_k - x_{k-1} \qquad (k = 2, 3, \ldots).$$

Solving for x_k in terms of x_1, we have

$$(4.5.6) \qquad x_k = kx_1 - \frac{k(k-1)}{2}\cdot\frac{c_1}{c_2} \quad (k = 2, 3, \ldots).$$

x_k must be positive, from which we have $x_1 > [(k-1)/2](c_1/c_2)$, implying that we make only a finite number of inspections, say n. Since $x_k = T$, we have

$$(4.5.7) \qquad x_k = kT/n + k(n-k)(c_1/2c_2) \quad (k = 0, 1, 2, \ldots).$$

Note that

$$(4.5.8) \qquad \delta_{n-1} = x_n - x_{n-1} = T/n - (c_1/2c_2)(n - 1) > 0,$$

i.e.,

$$(4.5.9) \qquad n(n-1) < 2c_2T/c_1.$$

But the expected cost using $n+1$ inspections minus the expected cost using n inspections is

$$(4.5.10) \quad -(c_2/2T)[(n+1)/n][T/(n+1) - (c_1n)/2c_2]^2 < 0.$$

Hence n must be the largest integer satisfying (4.5.9). Specifying n, we can determine the optimal time sequence $\{x_k\}$ $(x_1 < x_2 < \ldots < x_n)$ from (4.5.7).

As a numerical example, we assume that $T = 100$, $c_1 = 2$, and $c_2 = 1$. Then the optimal $n = 10$ from (4.5.9), and the optimal inspection sequences are {19, 36, 51, 64, 75, 84, 91, 96, 99, 100} from $x_k = k(20-k)$ $(k = 1, 2, \ldots, 10)$.

The algorithm described above is not easy to obtain the optimal inspection sequences $\{x_k\}$ in general except the simplest cases. We now develop an approximation model by introducing the inspection density $n(t)$ at time t, which is a smooth function denoting the number of inspections at time t. Introducing the inspection density $n(t)$, we now obtain the expected cost $C_0(n(t))$: The expected inspection cost to the detection of failure is

(4.5.11) $$c_1\int_0^\infty \int_0^t n(s)dsdF(t) = c_1\int_0^\infty n(t)\bar{F}(t)dt,$$

and the expected cost suffered for the system down for the time elapsed between system failure and its detection is approximated by

(4.5.12) $$c_2 \int_0^\infty [1/2n(t)]dF(t).$$

Then the approximated total expected cost to the detection of failure is

(4.5.13) $$C_0(n(t)) = c_1\int_0^\infty n(t)\bar{F}(t)dt + c_2\int_0^\infty [1/2n(t)]dF(t),$$

where $\bar{F}(t) \equiv 1 - F(t)$. Our objective is to obtain a unknown function $n(t)$ for which $C_0(n(t))$ is minimized. This is a typical variational problem with an unknown function $n(t)$. Euler's equation for the variational problem is given by

(4.5.14) $$c_2/[n(t)]^2 = 2c_1/r(t)$$

where $r(t) = f(t)/\bar{F}(t)$, the failure rate. Solving (4.5.14) in terms of $n(t)$, we have

(4.5.15) $$n(t) = \sqrt{(c_2/2c_1)r(t)}.$$

It is noted that Euler's equation (4.5.14) is a necessary condition. However, we can prove that the condition (4.5.14) is also a sufficient condition for the variational problem. Once $n(t)$ is given, we can easily obtain the optimal inspection sequence by

(4.5.16) $$1 = \int_{x_k}^{x_{k+1}} n(t)dt \quad (k = 0, 1, 2, \dots; x_0 = 0),$$

or

$$(4.5.17) \qquad k = \int_0^{x_k} n(t)dt \qquad (k = 1, 2, 3, \ldots).$$

Fig. 4.5.1 shows how to calcuate the optimal inspection sequences $\{x_k\}$, where each sectional area is a unity. From Fig. 4.5.1, we can also show that, if $F(t)$ is IFR (i.e., $r(t)$ is increasing), the optimal inspection intervals are nonincreasing, which proved in a theorem by Barlow and Proschan (1965).

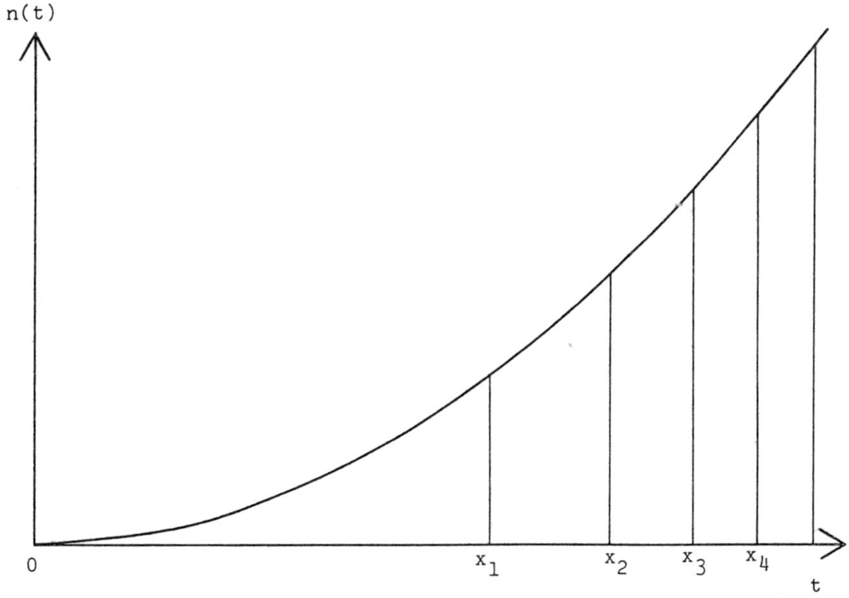

Fig. 4.5.1. A graph showing how to calculate the optimal inspection sequence $\{x_k\}$.

(4.5.18) Example We consider the same example (4.5.4) by the variational approach. Noting that

$$r(t) = f(t)/\bar{F}(t) = 1/(T-t), \tag{4.5.19}$$

we have

$$n(t) = \sqrt{(c_2/2c_1)}/\sqrt{T-t} \tag{4.5.20}$$

and

$$k = \sqrt{(c_2/2c_1)} \int_0^{t_k} dt/\sqrt{T-t} = \sqrt{2c_2/c_1}(\sqrt{T} - \sqrt{T-x_k}). \tag{4.5.21}$$

Solving for t_k implies

$$x_k = k(\sqrt{(2c_1T/c_2)} - kc_1/2c_2) \quad (k = 1, 2, \ldots). \tag{4.5.22}$$

Consider the same numerical example in which $T = 100$, $c_1 = 2$ and $c_2 = 1$. Then

$$x_k = k(20 - k) \quad (k = 1, 2, \ldots, 10), \tag{4.5.23}$$

which is just the same of the optimal inspection sequence obtained in Example (4.5.4).

(4.5.24) Example If we assume that $F(t)$ is governed by the Weibull distribution in (1.4.20). Then

$$r(t) = m\lambda^m t^{m-1}, \tag{4.5.25}$$

and from (4.5.17), we have

$$x_k = [c_1/(2m\lambda^m c_2)]^{1/(m+1)}[k(m+1)]^{2/(m+1)}. \tag{4.5.26}$$

In particular, if $m = 1$,

(4.5.27) $$x_k = 2k[c_1/(2\lambda c_2)]^{1/2} \quad (k = 1, 2, \ldots),$$

and

(4.5.28) $$\delta_k = x_k - x_{k-1} = 2[c_1/(2\lambda c_2)]^{1/2} \quad (k = 1, 2, \ldots).$$

It is clear from equation (4.5.26) that the optimal inspection intervals are increasing when $0 < m < 1$, constant when $m = 1$, and decreasing when $m > 1$. Of course, this fact can be easily verified by noting equation (4.5.17) and the IFR, CFR, and DFR properties of the failure rate $r(t)$.

In the first part of this section we have shown the assumptions for the basic inspection model. We propose a modified model by applying the variational approach just described above. For the basic model, we have assumed that each inspection is instantaneous. However, we eliminate this assumption and further assume the following: Each inspection is executed with a finite checking time T_c, which is the delay between the inspection and the recognition of its result. Other assumptions are just same as to the basic inspection model.

Then, the total expected cost is approximated by

(4.5.29) $$C_d(n(t)) = c_1 \int_0^\infty n(t)\bar{F}(t-T_c)dt + c_2[\int_0^\infty 1/[2n(t)]dF(t) + T_c].$$

Euler's equation implies

(4.5.30) $$c_1\bar{F}(t-T_c) - c_2 f(t)/[2n^2(t)] = 0.$$

Solving for $n(t)$ yields

(4.5.31) $$n(t) = \{[c_2/(2c_1)]r(t)\bar{F}(t)/\bar{F}(t-T_c)\}^{1/2}.$$

Substituting $n(t)$ in equation (4.5.31) into equation(4.5.17), we can obtain the optimal inspection sequence $\{x_k\}$, which is similar to the basic inspection model.

(4.5.32) Example If we assume the exponential lifetime distribution $F(t) = 1 - \exp(-\lambda t)$ $(t \geq 0, \lambda > 0)$, then we have

(4.5.33) $$n(t) = [(c_2/2c_1)\exp\{\max(0, t-T_c) - t\}]^{1/2}$$

$$= \begin{cases} [(c_2/2c_1)\exp(-t)]^{1/2} & (t \leq T_c) \\ [(c_2/2c_1)\exp(-T_c)]^{1/2} & (t \geq T_c). \end{cases}$$

Thus, if $t \geq T_c$, then

(4.5.34) $$x_k - x_{k-1} = 1/[(c_2/2c_1)\exp(-T_c)]^{1/2}.$$

That is, the inspection intervals increase initially and are constant after that.

For the basic inspection model, we assume that each inspection is perfect, i.e., the system failure can be detected with probability 1 by an inspection followed by the system failure. Here we eliminate this assumption and further assume that the following: The system failure can be detected with probability $1-b$ and cannot be detected with probability b, even though the system failure takes place, where $0 \leq b \leq 1$. That is, the system failure can be detected by the Bernoulli trials with success probability 1-b. Note that the Bernoulli trials for detecting the system failure follows a geometric distribution with respective parameters $p = 1-b$ and $q = b$ (see Table 1.2.1).

Then, the total expected cost is approximated by

$$(4.5.35) \qquad C_p(n(t)) = c_1[\int_0^\infty n(t)\bar{F}(t)dt + b/(1-b)]$$

$$= c_2(1+b)/(1-b)\int_0^\infty 1/[2n(t)]dF(t).$$

Euler's equation implies

$$(4.5.36) \qquad c_1\bar{F}(t) - c_2(1+b)f(t)/[2(1-b)n^2(t)] = 0.$$

Solving for $n(t)$ yields

$$(4.5.37) \qquad n(t) = [\{(1+b)(c_2/2c_1)/(1-b)\}r(t)]^{1/2}.$$

Substituting $n(t)$ in equation (4.5.37) into equation (4.5.17) yields the optimal inspection sequence $\{x_k\}$, which is similar to the preceding two inspection models.

(4.5.38) Example If we assume that the lifetime obeys a Weibull distribution $F(t) = 1 - \exp[-(\lambda t)^m]$ $(t \geq 0, \lambda > 0, m > 0)$, then we have

$$(4.5.39) \qquad n(t) = [\{(1+b)(c_2/2c_1)/(1-b)\}m\lambda^m t^{m-1}]^{1/2},$$

which can yield the optimal inspection sequence $\{x_k\}$ from equation (4.5.17).

Bibliography and Comments

Section 4.1: We are following MIL-STD 721 B (1966) and IEC (International Electrotechnical Commission) Publication (1978) for basic terms, definitions and related mathematics for reliability. See also Barlow and Proschan (1965, 1975).

Section 4.2: We are following a paper by Baxter (1981).

Section 4.3: Replacement models were summarized by Barlow and Proschan (1965, 1975). For discrete time replacement models, we are following the papers by Nakagawa and Osaki (1977) and Nakagawa (1984a, 1984b). For block replacement models, we are following a paper by Nakagawa (1979).

Section 4.4: We are following the papers by Osaki (1977b), Nakagawa and Osaki (1978), Thomas and Osaki (1978a, 1978b), Osaki (1979), Osaki, Kaio and Yamada (1981), and Kaio and Osaki (1979).

Section 4.5: Basic inspection models were discussed by Barlow and Proschan (1965). Inspection models introducing the inspection density were discussed by Keller (1974), Osaki (1977a, 1980), and Kaio and Osaki (1984, 1985).

[1] R.E. Barlow and F. Proschan (1965), Mathematical Theory of Reliability, Wiley, New York.

[2] R.E. Barlow and F. Proschan (1975), Statistical Theory of Reliability and Life Testing: Probability Models, Holt, Rinehart and Winston, New York.

[3] L.A. Baxter (1981), "Availability Measures for a Two-State System," J. Appl. Probability, Vol. 18, pp.227-235.

[4] International Electrotechnical Commission (1978), List of Basic Terms, Definitions and Related Mathematics for Reliability, Publication 271A, Bureau Central de la Commission Electrotechnique International, Genève.

[5] N. Kaio and S. Osaki (1979), "Extended Optimum Ordering Policies," (in Japanese), Trans. IECE, Vol. 62-A, pp. 373-380.

[6] N. Kaio and S. Osaki (1984), "Some Remarks on Optimum Inspection Policies," IEEE Trans. Reliability, Vol. R-33, pp. 277-279.

[7] N. Kaio and S. Osaki (1985), "Optimal Inspection Policies: A Review and Comparison," unpublished paper.

[8] J.B. Keller (1974), "Optimum Checking Schedules for Systems Subject to Random Failure," Management Sci., Vol. 21, pp. 256-260.

[9] Military Standard (1966), Definitions of Effectiveness Terms for Reliability, Maintainability, Human Factors and Safety, MIL-STD-721B, Washington.

[10] T. Nakagawa (1979), "A Summary of Block Replacement Policies," R.A.I.R.O. (Recherche operationelle), Vol. 13, pp. 351-361.

[11] T. Nakagawa (1984), "A Summary of Discrete Replacement Policies," Eurp. J. Operational Res., Vol. 17, pp. 382-392.

[12] T. Nakagawa (1984), "Discrete Replacement Models," in Stochastic Models in Reliability Theory, S. Osaki and Y. Hatoyama (Eds.), Springer-Verlag, Berlin.

[13] T. Nakagawa and S. Osaki (1977), "Discrete Time Age Replacement Policies," Operational Res. Q., Vol. 28, pp. 881-885.

[14] T. Nakagawa and S. Osaki (1978), "Optimum Ordering Policies with Lead Time for an Operating Unit," R.A.I.R.O. (Recherche operationelle), Vol. 12, pp. 383-393.

[15] S. Osaki (1977a), "An Ordering Policy with Lead Time," International J. Systems Sci., Vol. 8, pp. 1091-1095.

[16] S. Osaki (1977b), "A Note on Optimum Checking Procedures," (in Japanese), Trans. IECE, Vol. 60-A, pp. 764-765.

[17] S. Osaki (1979), "An Ordering Policy with Discounting," (in Japanese), Trans. IECE, Vol. 62-A, pp. 29-35.

[18] S. Osaki (1980), "An Optimum Checking Policy with Delay," (in Japanese), Trans. IECE, Vol. 63-A, pp. 463-464.

[19] S. Osaki, N. Kaio and S. Yamada (1981), "A Summary of Optimal Ordering Policies," IEEE Trans. Reliability, Vol. R-30, pp. 272-277.

[20] L.C. Thomas and S. Osaki (1978a), "An Optimal Ordering Policy for a Spare Unit with Lead Time," Eurp. J. Operational Res., Vol. 2, pp. 409-419.

[21] L.C. Thomas and S. Osaki (1978b), "A Note on Ordering Policy," IEEE Trans. Reliability, Vol. R-27, pp. 380-381.

CHAPTER 5

STOCHASTIC MODELS FOR TWO-UNIT REDUNDANT SYSTEMS

5.1 Introduction

In the preceding chapter, we have discussed a one-unit system. In Section 4.2, we have developed the availability theory for one-unit systems which assumed up and down states alternately. In Sections 4.2 and 4.3, we have developed basic replacement and ordering policies, and derived optimal replacement and ordering policies under suitable assumptions and criteria. Stochastic models discussed in Chapter 4 have been assumed one-unit systems where the lifetime and repair time of each unit were given by arbitrary and known distributions.

It is generally true that the high reliability and availability can be achieved by redundancy and/or maintenance. We discuss a two-unit redundant system as one of the basic redundant systems since there are many applications of two-unit redundant systems in the real world. For instance, we can encounter many computing systems composed of two processors for performing the computational demands and achieving the high reliability and performance. Examples of such systems are a banking system, an electronic switching system, a seat reservation system, and so on. Such systems are typical two-unit redundant systems.

We should classify two-unit redundant systems into two categories: One is a two-unit parallel redundant system and another is a two-unit standby redundant system. A two-unit parallel redundant system is a system that both two units are operating if they are available. On the other hand, a two-unit standby redundant system is a system that one is operating as online and another unit is in standby as offline if the latter is available. In particular, for fault-tolerant computing systems which will be discussed in the following chapter, a two-unit parallel redundant system is referred to as a dual system, which is composed of two identical units, executes the same tasks on each identical unit and checks the output. On the other hand, a two-unit standby redundant system is referred to as a duplex system, which is composed of an active unit (online) and a spare unit (offline).

In Section 5.2, we develop stochastic models for a two-unit parallel redundant system, and derive the reliability and availability measures by applying Markov renewal processes with non-regeneration point(s). We also analyze a two-unit parallel redundant system with bivariate exponential lifetime. In Section 5.3, we develop stochastic models for a two-unit standby redundant system by applying

Markov renewal processes under the most generalized assumptions, and derive the similar measures above by applying the Markov renewal processes with non-regeneration points. Throughout this chapter we mainly apply Markov renewal processes developed in Chapter 3 to analyses.

5.2 Two-Unit Parallel Redundant Systems

Consider a two-unit parallel redundant system composed of two identical units. We assume that each lifetime is distributed exponentially and each repair time is distributed arbitrarily. Otherwise, we cannot explicitly analyze the system if both distributions are assumed to be arbitrary.

We make the following assumptions for a two-unit parallel redundant system:

(i) The system is composed of two identical units, where each lifetime is distributed exponentially and each repair time distribution is arbitrary $G(t)$ $(t \geq 0)$ with finite mean $1/\mu$.

(ii) The system has a single repair facility and the repair discipline is 'first come, first served.' Each unit recovers its functioning (i.e., as good as new) upon repair completion.

(iii) Each switchover is perfect and each switchover time is negligible.

(iv) Two units are operating if they are active.

Define the following system states: State j ($j = 0, 1, 2$) corresponds to the number of units under repair or waiting for repair. We make the assumption that the failure rate in state j ($j = 0, 1$) is λ_j. This assumption implies the fruitful results of special cases discussed later. For this assumption, we have never given the parameter of the exponential lifetime distribution for each unit.

Define the following time instants (states) at which the process makes a transition into states:

State 0: Two unit are operating.
State 1: One unit is operating and the repair of the other unit starts.
State 2: The operating unit fails while the other unit is under repair.

The states $i = 0, 1$ are regeneration points. However state 2 is a non-regeneration point since the repair time distribution is arbitrary and the behavior after state 2 depends on the elapsed time from state 1.

We have the following mass functions or one-step transition probabilities of the Markov renewal process under consideration (except from state 2):

(5.2.1) $Q_{01}(t) = 1 - \exp(-\lambda_0 t)$,

(5.2.2) $Q_{10}(t) = \int_0^t e^{-\lambda_1 t} dG(t)$,

$$(5.2.3) \quad Q_{12}(t) = \int_0^t \lambda_1 e^{-\lambda_1 t}\bar{G}(t)dt,$$

where $\bar{G}(t) \equiv 1 - G(t)$.

However, it is impossible to obtain the one-step transition probability $Q_{21}(t)$ explicitly since state 2 is not a regeneration point and $Q_{21}(t)$ depends on the history of how long the repair time elapses. Define the two-step transition probability $Q_{11}^{(2)}(t)$ from state 1 to state 1 via state 2:

$$(5.2.4) \quad Q_{11}^{(2)}(t) = \int_0^t (1 - e^{-\lambda_1 t})dG(t).$$

The Laplace-Stieltjes transforms of one-step and two-step transition probabilities are the following:

$$(5.2.5) \quad Q_{01}^*(s) = \lambda_0/(s + \lambda_0) ,$$

$$(5.2.6) \quad Q_{10}^*(s) = G^*(s+\lambda_1) ,$$

$$(5.2.7) \quad Q_{12}^*(s) = [\lambda_1/(s + \lambda_1)][1 - G^*(s+\lambda_1)] ,$$

$$(5.2.8) \quad Q_{11}^{(2)*}(s) = G^*(s) - G^*(s+\lambda_1) ,$$

where $G^*(s)$ is the Laplace-Stieltjes transform of $G(t)$. Fig. 5.2.1 shows a signal-flow graph for the Markov renewal process, where the numbers circled are regeneration points and the number squared is a non-regeneration point. Being careful to state 2 which is a non-regeneration point, we can similarly apply the signal-flow graph approach discussed in Section 3.5.

Before discussing the first passage time distribution,

we have to notice the notation: Since we have used $G(t)$ as the repair time distribution, we define the first passage time distribution $H_{ij}(t)$ from state i $(i = 0, 1)$ to state j $(j = 0, 1, 2)$, instead of $G_{ij}(t)$ in Chapter 3.

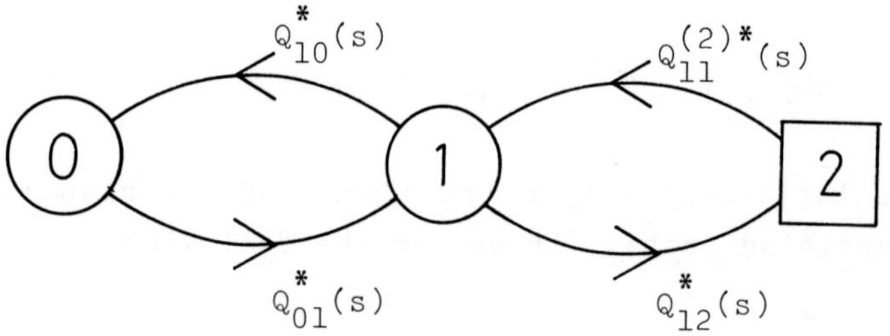

Fig. 5.2.1. A signal-flow graph for a two-unit parallel redundant system.

Derive the Laplace-Stieltjes transform $H^*_{i2}(s)$ $(j = 0, 1)$ of $H_{i2}(t)$. Noting that state i $(i = 0, 1)$ is a source and state 2 is a sink, we have the following results by applying Mason's gain formulae in the signal-flow graph on Fig. 5.2.1:

(5.2.9) $$H^*_{12}(s) = Q^*_{02}(s)/[1 - Q^*_{01}(s)Q^*_{10}(s)],$$

(5.2.10) $$H^*_{02}(s) = Q^*_{01}(s)\ H^*_{12}(s)$$

$$= Q^*_{01}(s)Q^*_{12}(s)/[1 - Q^*_{01}(s)Q^*_{10}(s)].$$

The mean first passage time from state 0 to system failure, ℓ_{02}, is given in terms of the Laplace-Stieltjes transform $G^*(s)$:

(5.2.11) $$\ell_{02} = 1/\lambda_1 + 1/\{\lambda_0[1 - G^*(\lambda_1)]\}.$$

We can also obtain the following results:

(5.2.13) $$H^*_{11}(s) = Q^*_{10}(s)Q^*_{01}(s) + Q^{(2)*}_{11}(s),$$

(5.2.14) $$\ell_{11} = 1/\mu + G^*(\lambda_1)/\lambda_0,$$

where $1/\mu = \int_0^\infty t dG(t)$, the mean repair time.

Recall that $M_{ij}(t)$ denotes the mean number of visits to state j $(j = 0, 1, 2)$ during the interval $(0, t]$, starting from state i $(i = 0, 1)$. To obtain the Laplace-Stieltjes transform $M^*_{ij}(s)$ of $M_{ij}(t)$, we can first show a signal-flow graph in Fig. 5.2.2 for $M^*_{i1}(s)$ $(i = 0, 1)$, the techniques of which have been discussed in Section 3.3. From Fig. 5.2.2, we have the following results:

$$(5.2.15)\quad M_{11}^{*}(s) = [Q_{10}^{*}(s)Q_{01}^{*}(s) + Q_{11}^{(2)*}(s)]/[1 - Q_{10}^{*}(s)Q_{01}^{*}(s) - Q_{11}^{(2)*}(s)] = H_{11}^{*}(s)/[1 - H_{11}^{*}(s)] ,$$

$$(5.2.16)\quad M_{01}^{*}(s) = Q_{01}^{*}(s)[1 + M_{11}^{*}(s)].$$

Just similar to the above techniques, we have

$$(5.2.17)\quad M_{1j}^{*}(s) = Q_{1j}^{*}(s)/[1 - H_{11}^{*}(s)] \quad (j = 0, 2),$$

$$(5.2.18)\quad M_{00}^{*}(s) = Q_{01}^{*}(s)Q_{10}^{*}(s)/[1 - H_{11}^{*}(1)],$$

$$(5.2.19)\quad M_{02}^{*}(s) = Q_{01}^{*}(s)Q_{12}^{*}(s)/[1 - H_{11}^{*}(s)].$$

Note that, if we obtain the branch gain $1 \longrightarrow 2$ or $1 \longrightarrow 2 \longrightarrow 1$ in Fig. 5.2.2 as a path or a loop, we should consider $Q_{12}^{*}(s)$ or $Q_{11}^{(2)*}(s)$, respectively, since state 2 is not a regeneration point. We have used these techniques for obtaining $M_{ij}^{*}(s)$ in (5.2.15) - (5.2.19).

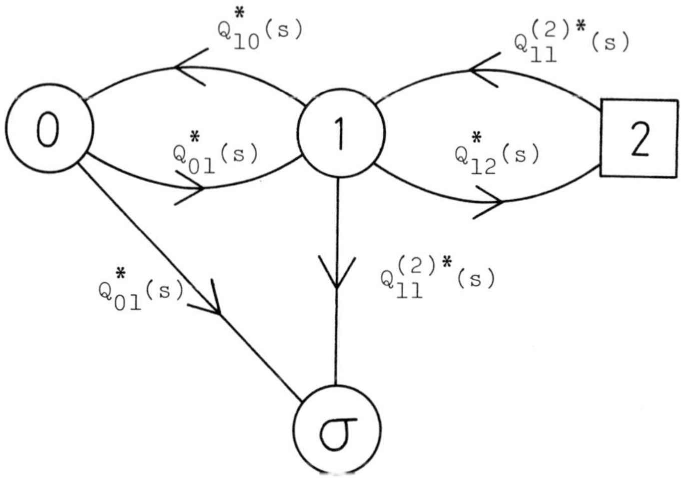

Fig. 5.2.2. A signal-flow graph for obtaining $M^*_{i1}(s)$ (i = 0, 1).

Further, $M_j = \lim_{t\to\infty} M_{ij}(t)/t$ (i = 0, 1; j = 0, 1, 2), the expected number of visits to state j per unit time in the steady-state, is given by

(5.2.20) $$M_0 = G^*(\lambda_1)/\ell_{11} ,$$

(5.2.21) $$M_1 = 1/\ell_{11} ,$$

(5.2.22) $$M_2 = [1 - G^*(\lambda_1)]/\ell_{11} ,$$

where ℓ_{11} is given in (5.2.14). Note that M_2 represents the mean number of system failures per unit time in the steady-state.

Recall that $P_{ij}(t)$ denotes the transition probability that the process is in state j ($j = 0, 1, 2$) at time t starting from state i ($i = 0, 1$) at time 0. The Laplace-Stieltjes transform $P^*_{ij}(s)$ of $P_{ij}(t)$ can be similarly obtained by applying the signal-flow graph techniques developed in Section 5.3. However, it is also easy to derive $P^*_{ij}(s)$ intuitively. We have the following renewal-type equations for the Laplace-Stieltjes transform $P^*_{1j}(s)$ ($j = 0, 1, 2$):

(5.2.23) $$P^*_{10}(s) = Q^*_{10}(s)[1 - Q^*_{01}(s)] + H^*_{11}(s)P^*_{10}(s) ,$$

(5.2.24) $$P^*_{11}(s) = 1 - Q^*_{10}(s) - Q^*_{12}(s) + H^*_{11}(s)P^*_{11}(s) ,$$

(5.2.25) $$P^*_{12}(s) = Q^*_{12}(s) - Q^{(2)*}_{11}(s) + H^*_{11}(s)P^*_{12}(s) .$$

For $P^*_{0j}(s)$ ($j = 0, 1, 2$), using $P^*_{1j}(s)$ ($j = 0, 1, 2$), we have

(5.2.26) $$P^*_{00}(s) = 1 - Q^*_{01}(s) + Q^*_{01}(s)P^*_{10}(s) ,$$

(5.2.27) $$P^*_{0j}(s) = Q^*_{01}(s)P^*_{1j}(s) \quad (j = 1, 2) .$$

Solving for $P^*_{0j}(s)$ ($j = 0, 1, 2$), we have

$$(5.2.28)\quad P_{00}^{*}(s) = [1 - Q_{01}^{*}(s)][1 - Q_{01}^{*}(s)Q_{10}^{*}(s)]/[1 - H_{11}^{*}(s)],$$

$$(5.2.29)\quad P_{01}^{*}(s) = Q_{01}^{*}(s)[1 - Q_{10}^{*}(s) - Q_{12}^{*}(s)]/[1 - H_{11}^{*}(s)],$$

$$(5.2.30)\quad P_{02}^{*}(s) = Q_{01}^{*}(s)[Q_{12}^{*}(s) - Q_{11}^{(2)*}(s)]/[1 - H_{11}^{*}(s)],$$

Of course, we can derive $P_{ij}^{*}(s)$ $(i = 0, 1; j = 0, 1, 2)$ in terms of λ_0, λ_1 and $G^{*}(s)$ by using the previous results. It can be easily verified that

$$(5.2.31)\quad P_{00}^{*}(s) + P_{01}^{*}(s) + P_{02}^{*}(s) = 1,$$

which implies that the total probability is a unity for any $t \geq 0$.

There exists the limiting probability $P_j = \lim_{t\to\infty} P_{ij}(t)$ $(i = 0, 1; j = 0, 1, 2)$ which is independent of the initial state i under the suitable conditions. We have

$$(5.2.32)\quad P_0 = 1 - 1/(\mu \ell_{11}),$$

$$(5.2.33)\quad P_1 = [1 - G^{*}(\lambda_1)]/(\lambda_1 \ell_{11}),$$

$$(5.2.34)\quad P_2 = 1/(\mu \ell_{11}) - [1 - G^{*}(\lambda_1)]/(\lambda_1 \ell_{11}).$$

Note that $P_{00}(t) + P_{01}(t)$ represents the pointwise (instantaneous) availability of the system if the process starts from state 0 and $P_0 + P_1$ represents the steady-state (or limiting) availability.

It is noted that the process repeats up and down states alternately, and the limiting availability is given by

(5.2.35) $$P_0 + P_1 = \text{MTBF}/(\text{MTBF} + \text{MDT}),$$

and the reciprocal of M_2, the mean number of visits to system failure per unit time in the steady-state, is

(5.2.36) $$1/M_2 = \text{MTBF} + \text{MDT},$$

which is the time duration of one cycle (up time plus down time). Thus, we have

(5.2.37) $$\text{MTBF} = (P_0 + P_1)/M_2,$$

(5.2.38) $$\text{MDT} = P_2/M_2.$$

In the preceding section, we have just classified into two categories (i.e., two-unit standby and parallel redundant systems) for a two-unit redundant system. However, introducing the concepts of cold, warm, and hot standby redundancies, we have three categories for a two-unit redundant system: That is, "cold standby" means that a standby unit neither deteriorates nor fails, "warm standby" means that a standby unit can fail, but the probability of standby failure might be less than or equal to that of operating failure at any time $t \geq 0$, and finally "hot standby" means that a standby unit can fail just same to an operating unit, which implies a two-unit parallel redundant system from the viewpoint of stochastic modeling.

The model discussed above includes several interesting models as special cases. We show three cases which are well-known and applicable in practice.

(i) $\lambda_0 = \lambda_1 \equiv \lambda$.

In this case, the failure rate λ is identical for states 0 and 1, which implies a two-unit "cold" standby redundant system. For this case, we have assumed the exponential lifetime $F(t) = 1 - \exp(-\lambda t)$ $(t \geq 0)$. However, we will develop the analysis under the assumptions that both $F(t)$ and $G(t)$ are arbitrary in the following section.

(ii) $\lambda_0 = 2\lambda$ and $\lambda_1 = \lambda$.

In this case, the failure rates are 2λ for state 0 and λ for state 1, which implies a two-unit parallel redundant system (i.e., a two-unit "hot" standby redundant system). Such a two-unit parallel redundant system has been discussed and variations of such a system have been discussed by many authors.

(iii) $\lambda_0 = \lambda + \lambda'$ and $\lambda_1 = \lambda$.

In this case, the failure rates are $\lambda + \lambda'$ for state 0 and λ for state 1, which implies that, if the lifetime of an operating unit is distributed exponentially with parameter λ and the lifetime of a standby unit is distributed exponentially with parameter λ' (of course, $\lambda \geq \lambda'$), this case corresponds to a two-unit "warm" standby redundant system.

There are many variations of a two-unit parallel redundant system. One of variations is to generalize a two-unit parallel redundant system of two dissimilar units. We further assume that the lifetimes of two dissimilar units are not independent, but are distributed by the bivariate exponential distribution whose survival probability is given by

(5.2.39) $$\bar{F}(x,y) = \exp[-\lambda_1 x - \lambda_2 y - \lambda_{12}\max(x,y)] ,$$

which has been given in (1.3.13), where the random variables X and Y denotes the lifetimes of units 1 and 2, respectively, and $x \geq 0$, $y \geq 0$, $\lambda_1 > 0$, $\lambda_2 > 0$ and $\lambda_{12} \geq 0$. Of course, $\lambda_{12} = 0$ implies that X and Y are independent and distributed exponentially with respective parameters λ_1 and λ_2. We assume that the repair time of units 1 and 2 are independent and distributed by arbitrary distributions $G_1(t)$ and $G_2(t)$ with finite means $1/\mu_1$ and $1/\mu_2$, respectively. The assumptions (ii), (iii) and (iv) for a two-unit parallel redundant system are just same.

Define the following time instants (states) at which the process makes a transition into states:

State 0: Two units are operating.

State 1: Upon failure of unit 1, the repair is made immediately while unit 2 is still operating.

State 2: Upon failure of unit 2, the repair is made immediately while unit 1 is still operating.

State 3: Upon simultaneous failures of units 1 and 2, the repair of unit 1 is made immediately (system failure).

State 4: Before the repair of unit 1 is completed, unit 2 fails (system failure).

State 5: Before the repair of unit 2 is completed, unit 1 fails (system failure).

It is generally true that states 0, 1, 2, and 3 are regeneration points and states 4 and 5 are non-regeneration points. The state transition diagram is shown in Fig. 5.2.3, where the number <u>circled</u> denotes a regeneration point and the number <u>squared</u> a non-regeneration point.

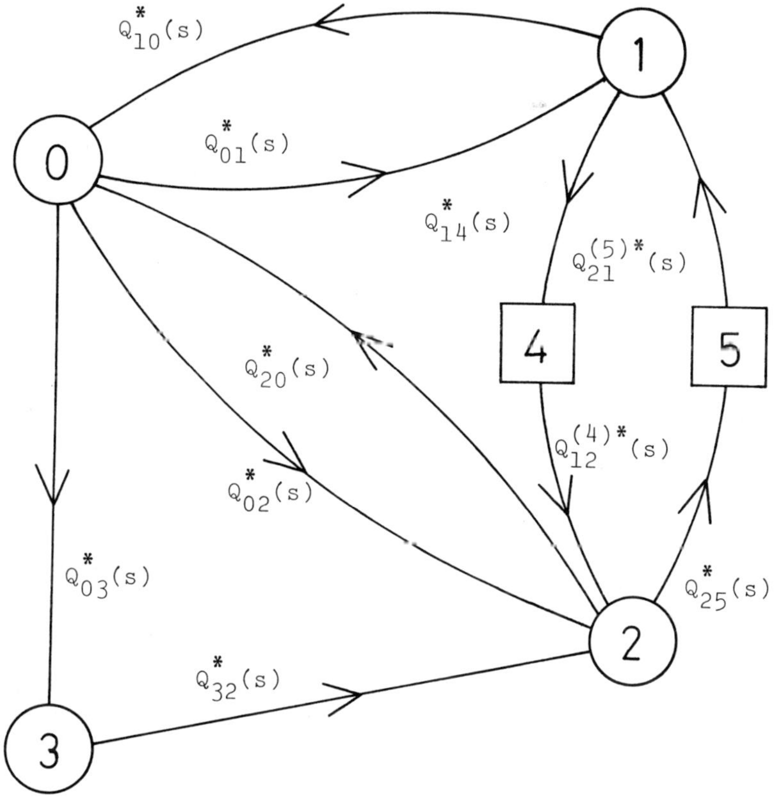

Fig. 5.2.3. A signal-flow graph for a two-unit parallel redundant system with bivariate exponential lifetimes.

The Laplace-Stieltjes transforms $Q^*_{ij}(s)$ and $Q^{(k)*}_{ij}(s)$ are the following:

$$(5.2.40) \quad Q_{0i}^{*}(s) = \lambda_i/(s + \Lambda) \quad (i = i, 2) ,$$

$$(5.2.41) \quad Q_{03}^{*}(s) = \lambda_{12}/(s + \Lambda) ,$$

$$(5.2.42) \quad Q_{i0}^{*}(s) = G_i^{*}(s + \lambda_{3-i} + \lambda_{12}) \quad (i = 1, 2) ,$$

$$(5.2.43) \quad Q_{32}^{*}(s) = G_1^{*}(s) ,$$

$$(5.2.44) \quad Q_{i,i+3}^{*}(s) = \frac{\lambda_{3-i} + \lambda_{12}}{s + \lambda_{3-i} + \lambda_{12}}[1 - G_i^{*}(s+\lambda_{3-i}+\lambda_{12})] \quad (i = 1, 2),$$

$$(5.2.45) \quad Q_{i,3-i}^{(i+3)*}(s) = G_i^{*}(s) - G_i^{*}(s + \lambda_{3-i} + \lambda_{12}) \quad (i = 1, 2),$$

where $\Lambda \equiv \lambda_1 + \lambda_2 + \lambda_{12}$ and $G_i^{*}(s)$ $(i = 1, 2)$ is the Laplace-Stieltjes transform of $G_i(t)$, respectively.

Noting that states 3, 4 and 5 denote the system failure, we obtain the Laplace-Stieltjes transform of the distribution to system failure starting from state 0:

$$(5.2.46) \quad H_{0F}^{*}(s) = \frac{Q_{03}^{*}(s) + Q_{01}^{*}(s)Q_{14}^{*}(s) + Q_{02}^{*}(s)Q_{25}^{*}(s)}{1 - Q_{01}^{*}(s)Q_{10}^{*}(s) - Q_{02}^{*}(s)Q_{20}^{*}(s)} ,$$

and the Mean Time to First Failure (MTFF):

$$(5.2.47) \quad \ell_{0F} = \frac{\xi_0 + Q_{01}^{*}(0)\xi_1 + Q_{02}^{*}(0)\dot{\xi}_2}{1 - Q_{01}^{*}(0)Q_{10}^{*}(0) - Q_{02}^{*}(0)Q_{20}^{*}(s)} ,$$

where ξ_0, ξ_1 and ξ_2 are the unconditional means:

$$(5.2.48) \quad \xi_0 = 1/\Lambda ,$$

$$(5.2.49) \quad \xi_i = [1 - G_i^*(\lambda_{3-i} + \lambda_{12})]/(\lambda_{3-i} + \lambda_{12}) \quad (i = 1,2),$$

First of all, we can obtain the Laplace-Stieltjes transform $H_{ii}(s)$ of the recurrence time distribution $(i = 0, 1, 2, 3)$ by using the signal-flow graph approach developed in Section 3.4. We can also obtain the mean recurrencce time ℓ_{ii} for state i $(i = 0, 1, 2, 3)$.

Using the above results $H_{ij}^*(s)$ and ℓ_{ii} for states i $(i = 0, 1, 2, 3)$, we obtain the Laplace-Stieltjes transform $M_{ij}^*(s)$ of the generalized renewal function $M_{ij}(t)$, i.e., the mean number of visits to state j $(j = 0, 1, 2, 3, 4, 5)$ during the interval $(0, t]$, starting from state i $(i = 0, 1, 2, 3)$ at time 0:

$$(5.2.50) \quad M_{ii}^*(s) = H_{ii}^*(s)/[1 - H_{ii}^*(s)] \quad (i = 0, 3) ,$$

$$(5.2.51) \quad M_{ii}^*(s) = H_{ii}^*(s)Q_{i0}^*(s)/[1 - H_{ii}^*(s)] \quad (i = 1, 2) ,$$

$$(5.2.52) \quad M_{i,i+3}^*(s) = H_{ii}^*(s)Q_{i,i+3}^*(s)/[1 - H_{ii}^*(s)] \quad (i = 1, 2) ,$$

which can be easily obtained by renewal-theoretic arguments. Of course, we can also obtain the same results by applying the signal-flow graph approach in Section 3.4. However, the results are very complicated since they are not expressed in terms of $H_{ii}(s)$ but in terms of $Q_{ij}(s)$ and $Q_{ij}^{(k)*}(s)$. The mean numbers of visits to state i $(i = 0, 1, 2, 3, 4, 5)$ per unit time in the steady-state are given by

$$(5.2.53) \quad M_i = 1/\ell_{ii} \quad (i = 0, 3) ,$$

$$(5.2.54) \quad M_i = Q_{i0}^*(0)/\ell_{ii} \quad (i = 1, 2) ,$$

(5.2.55) $M_{i+3} = Q^*_{i,i+3}(0)/\ell_{ii}$ $(i = 1, 2)$.

We can obtain the Laplace-Stieltjes transform $P^*_{ij}(s)$ of the transition probability $P_{ij}(t)$ from state i to state j. However, we omit the results. In particular, the limiting probabilities $P_j = \lim_{t\to\infty} P_{ij}(t)$ are expressed in terms of ℓ_{ii} $(i = 0, 1, 2, 3)$ and the unconditional means η_i for state i $(i = 0, 1, 2, 3, 4, 5)$, not neglecting the non-regeneration points (i.e., states 4 and 5):

(5.2.56) $P_i = \eta_i/\ell_{ii}$ $(i = 0, 1, 2, 3)$,

(5.2.57) $P_4 = \eta_4/\ell_{11}$,

(5.2.58) $P_5 = \eta_5/\ell_{22}$,

where $\eta_i = \xi_i$ $(i = 0, 1, 2)$, which are given in equations (5.2.49) - (5.2.50), and

(5.2.59) $\eta_3 = 1/\mu_1$,

(5.2.60) $\eta_{i+3} = 1/\mu_i - [1 - G^*_i(\lambda_{3-i}+\lambda_{12})]/(\lambda_{3-i} + \lambda_2)$ $(i = 1, 2)$.

It is finally noted that the Laplace-Stieltjes transform of the pointwise availability starting from state 0 is $P^*_{00}(s) + P^*_{01}(s) + P^*_{02}(s)$ and the limiting availability is $P_0 + P_1 + P_2$. The mean number of visits to system failure per unit time in the steady-state is $M_3 + M_4 + M_5$.

It is finally noted that

(5.2.61) $$MTBF = (P_0 + P_1 + P_2)/(M_3 + M_4 + M_5) ,$$

(5.2.62) $$MDT = (P_3 + P_4 + P_5)/(M_3 + M_4 + M_5) ,$$

which can be similarly interpreted as shown in equations (5.2.35) - (5.2.38).

5.3 Two-Unit Standby Redundant Systems

Consider a two-unit standby redundant system composed of two identical units, where "standby" means "cold standby". We assume in this section that each lifetime and repair time are distributed arbitrarily. We discuss a two-unit standby redundant system under the most generalized assumptions above.

We also make the following assumptions for a two-unit standby redundant system:

(i) The system is composed of two identical units, where each lifetime distribution is an arbitrary $F(t)$ $(t \geq 0)$ with finite mean $1/\lambda$.

(ii) The system has a single repair facility and the repair discipline is 'first come, first served'. Each unit recovers its functioning (i.e., as good as new) upon repair completion. Each unit in standby neither deteriorates nor fails.

(iii) Each switchover is perfect and each switchover time is negligible.

(iv) One unit is operating and another unit is in standby if they are active.

Define the following time instants (states) at which the process makes a transition into states:

State -1: One unit begins operating and another unit is in standby.

State 0: One unit under repair completes its repair while another unit is operating.

State 1: One operating unit fails (and its repair immediately begins) while another unit is in standby.

State 2: One operating unit fails while another unit is under repair (system failure).

Note that state -1 is introduced for the initial starting time instant, and once the process moves from state -1 to state 1, the process moves among states 0, 1, and 2. From the definition above, it is clear that state 1 is a regeneration point and states 0 and 2 are non-regeneration points.

We have the following mass functions or one-step transition probabilities of the Markov renewal process under consideration (except from states 0 and 2):

(5.3.1) $$Q_{-11}(t) = F(t) ,$$

(5.3.2) $$Q_{10}(t) = \int_0^t \bar{F}(t)dG(t) ,$$

$$Q_{12}(t) = \int_0^t \bar{G}(t)dF(t) \ , \tag{5.3.3}$$

where $\bar{F}(t) \equiv 1 - F(t)$ and $\bar{G}(t) \equiv 1 - G(t)$.

However, it is impossible to obtain the one-step transition probabilities $Q_{10}(t)$ and $Q_{20}(t)$ explicitly since states 0 and 2 are non-regeneration points. Just similar to the discussion in Section 5.1, we define the two-step transition probabilities $Q_{11}^{(k)}(t)$ $(k = 0, 2)$ from state 1 to state 1 via state 2:

$$Q_{11}^{(0)}(t) = \int_0^t G(t)dF(t) \ , \tag{5.3.4}$$

$$Q_{11}^{(2)}(t) = \int_0^t F(t)dG(t) \ . \tag{5.3.5}$$

The Laplace-Stieltjes transforms of one-step and two-step transition probabilities are the following:

$$Q_{-11}^{*}(s) = F^{*}(s) \ , \tag{5.3.6}$$

$$Q_{10}^{*}(s) = \int_0^\infty e^{-st}\bar{F}(t)dG(t) \ , \tag{5.3.7}$$

$$Q_{12}^{*}(s) = \int_0^\infty e^{-st}\bar{G}(t)dF(t) \ , \tag{5.3.8}$$

$$Q_{11}^{(0)*}(s) = \int_0^\infty e^{-st}G(t)dF(t) \ , \tag{5.3.9}$$

$$Q_{11}^{(2)*}(s) = \int_0^\infty e^{-st}F(t)dG(t) \ , \tag{5.3.10}$$

where $F^{*}(s)$ is the Laplace-Stieltjes transform of $F(t)$. Fig. 5.3.1 shows a signal-flow graph for the Markov renewal process, where states 0 and 2 are non-regeneration points.

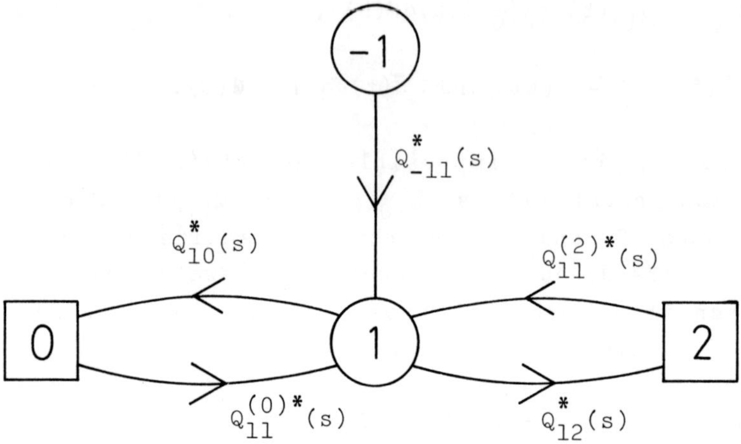

Fig. 5.3.1. A signal-flow graph for a two-unit standby redundant system.

Referring to Fig. 5.3.1, we obtain the recurrence time distribution $H_{11}(t)$ for state 1 which is a regeneration point:

$$H_{11}(t) = Q^{(0)}_{11}(t) + Q^{(2)}_{11}(t) = F(t)G(t) , \tag{5.3.11}$$

which is the distribution of the random variable of the maximum of two random variables (see equation (1.5.4)), and the process comes back to state 1 via state 0 or 2, whichever occurs first. The Laplace-Stieltjes transform $H^*_{11}(s)$ of $H_{11}(t)$ is given by

(5.3.12) $$H^*_{11}(s) = \int_0^\infty e^{-st} d[F(t)G(t)] ,$$

and the mean recurrence time ℓ_{11} for state 1 is given by

(5.3.14) $$\ell_{11} = \int_0^\infty t dH_{11}(t) = 1/\lambda + 1/\mu - 1/\gamma ,$$

where $1/\gamma \equiv \int_0^\infty \bar{F}(t)\bar{G}(t)dt$.

Consider the first passage time distribution $H_{i2}(t)$ from state i $(i = 0, 1)$ to state 2, i.e., the first passage time distribution to system failure. Referring to Fig. 5.3.1, we have the following Laplace-Stieltjes transforms $H^*_{i2}(s)$ $(i = 1, 2)$ of $H_{i2}(t)$:

(5.3.15) $$H^*_{12}(s) = Q^*_{12}(s)/[1 - Q^{(0)*}_{11}(s)] ,$$

(5.3.16) $$H^*_{-12}(s) = Q^*_{-11}(s)H^*_{12}(s) .$$

The mean first passage times from state i $(i = -1, 1)$ to system failure are

(5.3.17) $$\ell_{12} = 1/[\lambda(1 - Q^{(0)*}_{11}(0)] ,$$

(5.3.18) $$\ell_{-12} = 1/\lambda + 1/[\lambda(1 - Q^{(0)*}_{11}(0))] .$$

Let us derive the Laplace-Stieltjes transform $M^*_{ij}(s)$ of $M_{ij}(t)$, the mean number of visits to state j $(j = 0, 1, 2)$ during the interval $(0, t]$, starting from state i $(i = -1, 0)$. Just similar to the discussions in the preceding section, we have

$$(5.3.19)\qquad M_{11}^*(s) = H_{11}^*(s)/[1 - H_{11}^*(s)] ,$$

$$(5.3.20)\qquad M_{1j}^*(s) = Q_{1j}^*(s)/[1 - H_{11}^*(s)] \quad (j = 0, 2) ,$$

$$(5.3.21)\qquad M_{-1j}^*(s) = Q_{-11}^*(s)M_{1j}^*(s) \quad (j = 0, 2) ,$$

$$(5.3.22)\qquad M_{-11}^*(s) = Q_{-11}^*(s)[1 + M_{11}^*(s)] ,$$

$$= Q_{-11}^*(s)/[1 - H_{11}^*(s)] .$$

The expected numbers of visits to state j per unit time in the steady-state, $M_j = \lim_{t\to\infty} M_{ij}(t)$ $(i = -1, 1;\ j = 0, 1, 2)$, are given by

$$(5.3.23)\qquad M_0 = Q_{10}^*(0)/\ell_{11} ,$$

$$(5.3.24)\qquad M_1 = 1/\ell_{11} ,$$

$$(5.3.25)\qquad M_2 = Q_{12}^*(0)/\ell_{11} .$$

Note that M_2 represents the mean number of system failures per unit time in the steady-state.

Let us derive the transition probability $P_{ij}(t)$ from state i $(i = -1, 1)$. Noting that state 1 is only a regeneration point, we have the following Laplace-Stieltjes transforms:

$$(5.3.26)\quad P_{11}^*(s) = [1 - Q_{10}^*(s) - Q_{12}^*(s)]/[1 - H_{11}^*(s)],$$

$$(5.3.27)\quad P_{1j}^*(s) = [Q_{1j}^*(s) - Q_{11}^*(s)]/[1 - H_{11}^*(s)] \qquad (j = 0, 2),$$

$$(5.3.27)\quad P_{-1j}^*(s) = Q_{01}^*(s)P_{1j}^*(s) \qquad (j = 0, 1, 2).$$

From a Tauberian theorem (see Appendix A), we have the limitng probability $P_j = \lim_{s \to 0} P^*_{ij}(s)$ which is independent of the initial state i under the suitable conditions:

$$P_0 = 1 - 1/(\mu \ell_{11}) , \tag{5.3.29}$$

$$P_1 = 1/(\lambda \ell_{11}) + 1/(\mu \ell_{11}) - 1 , \tag{5.3.30}$$

$$P_2 = 1 - 1/(\lambda \ell_{11}) , \tag{5.3.31}$$

where it is evident that

$$P_1 + P_2 + P_3 = 1. \tag{5.3.32}$$

It is noted that the Laplace-Stieltjes transform of the pointwise availability is $P^*_{-10}(s) + P^*_{-11}(s)$ and the steady-state (or limiting) availability is $P_0 + P_1$. Note also that M_2 represents the mean number of visits to system failure per unit time in the steady-state.

It is finally noted that

$$\text{MTBF} = (P_0 + P_1)/M_2, \tag{5.3.33}$$

$$\text{MDT} = P_2/M_2, \tag{5.3.34}$$

which can be similarly interpreted as shown in equations (5.2.35) - (5.2.38).

As shown above, the two-unit standby redundant system assumes up and down states alternately and the uptime and downtime are dependent each other because state 2 is no longer a regeneration point. That is, the downtime depends on the history of how long the repair time elapses. Note that state -1 has been introduced to define the initial state that two unit are as good as new. Once the process moves to state 1, the process can move among states 0, 1, 2. In the sequel of this section, we are just interested in the behavior among states 0, 1, and 2. That is, the process starts in state 1 at time 0. A sample function of such a behavior is shown in Fig. 5.3.2.

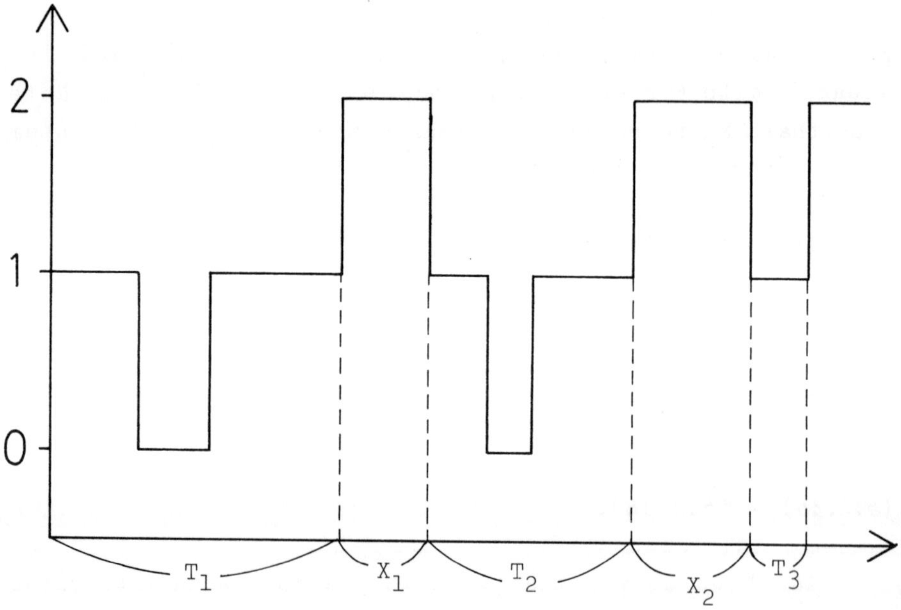

Fig. 5.3.2. A sample function of the behavior of uptime and downtime.

Let T and X denote the random variables of the single uptime and downtime, respectively. It is noted that the system is 'up' when one of the two units is at least up, and the system is 'down' when both two units are down simultaneously. Define the joint probability:

$$(5.3.35)\quad \Pi(t,x) \equiv P\{T \leq t \text{ and } X > x \mid \text{the process starts from state 1 at time 0}\}.$$

Let T_k and X_k $(k = 1, 2, \ldots)$ are the random variables of the kth uptime and k^{th} downtime, respectively (see Fig. 5.3.2). Once the process assumes up and down states alternately according to the random variables T_k and X_k $(k = 1, 2, \ldots)$, respectively, we have

$$(5.3.36)\quad \Pi(t, x) = P\{T_k \leq t \text{ and } X_k > x\} \quad (k = 1, 2, \ldots),$$

which specifies the behavior of the process completely since state 1 (i.e., the time instant at which the system recovery occurs) is a regeneration point. Then we have the following equation of renewal type:

$$(5.3.37)\quad \Pi(t, x) = \int_0^t \bar{G}(x+u)dF(u) + \int_0^t \Pi(t-u,x)G(u)dF(u).$$

The first term of the right-hand side of (5.3.37) is the probability that an operating unit fails during $(u, u+du]$ $(0 < u < t)$ and the repair of the failed unit is not completed up to time $x + u$. The second term is the probability that after repair completion of a failed unit, the operating unit fails during $(u, u+du]$, and then the process obey $\Pi(t-u, x)$.

Taking the Laplace-Stieltjes transforms on both sides of (5.3.37) with respect to t, we have

(5.3.38) $\Pi^*(s, x)$

$$= \int_0^\infty e^{-st}\bar{G}(x+t)dF(t)/[1 - \int_0^\infty e^{-st}G(t)dF(t)].$$

In particular, $\Pi^*(s, 0)$ denotes the Laplace-Stieltjes transform of the first passage time distribution from state 1 to state 2 (system failure) and is equal to $H^*_{12}(s)$ in (5.3.15). The survival probability of the down time, $\Pi(x) = P\{X > x\}$, is given by the marginal distribution of $\Pi(t, x)$, i.e.,

(5.3.39) $$\Pi(x) = \Pi(0, x) = \int_0^\infty \bar{G}(x+t)dF(t)/ \int_0^\infty \bar{G}(t)dF(t),$$

and the mean downtime is

(5.3.40) $$\int_0^\infty \Pi(x)dx = \int_0^\infty \bar{G}(t)F(t)dt/ \int_0^\infty \bar{G}(t)dF(t),$$

which has been given in (5.3.34).

In particular, if $G(t) = 1 - \exp(-\mu t)$, the uptime and downtime are independent each other, and it is evident that $\Pi(s, x) = \Pi(s, 0)\cdot\Pi(0, x)$.

Bibliography and Comments

Section 5.2: We are following the papers by Nakagawa and Osaki (1975) and Osaki (1980).

Section 5.3: We are following the papers by Nakagawa and Osaki (1974a, 1974b, 1976a). Osaki (1970), Osaki and Asakura (1970), and Nakagawa and Osaki (1974c, 1974d, 1976b) discussed preventive maintenance policies for a two-unit standby redundant system. However, we omit their results. The papers reviewing maintenance policies for stochastic systems are given by Osaki and Nakagawa (1976), Pierskalla and Voelker (1976), and Sherif and Smith (1981).

[1] T. Nakagawa and S. Osaki (1974a), "Stochastic Behaviour of a Two-Unit Standby Redundant System," INFOR (Canadian J. of Operational Research and Information Processing), Vol. 12, pp. 66-70.

[2] T. Nakagawa and S. Osaki (1974b), "Stochastic Behavior of a Two-Dissimilar-Unit Standby Redundant System with Repair Maintenance," Microelectron. Reliab., Vol. 13, pp. 143-148.

[3] T. Nakagawa and S. Osaki (1974c), "Optimum Preventive Maintenance Policies for a 2-Unit Redundant System," IEEE Trans. Reliability, Vol. R-23, pp. 86-91.

[4] T. Nakagawa and S. Osaki (1974d), "Optimum Preventive Maintenance Policies Maximizing the Mean Time to the First System Failure for a Two-Unit Standby Redundant System," J. of Optimization Theory and Applications, Vol. 14, pp. 115-129.

[5] T. Nakagawa and S. Osaki (1975), "Stochastic Behavior of Two-Unit Paralleled Redundant Systems with Repair Maintenance," Microelectron. Reliab., Vol. 14, pp. 457-461.

[6] T. Nakagawa and S. Osaki (1976a), "Joint Distributions of Uptime and Downtime for Some Repairable Systems," J. Operations Res. Soc. Japan, Vol. 19, 209-216.

[7] T. Nakagawa and S. Osaki (1976b), "A Summary of Optimum Preventive Maintenance Policies for a Two-Unit Standby Redundant System," Zeitschrift fur Operations Research, Band 20, pp. 171-187.

[8] S. Osaki (1970), "System Reliability Analysis by Markov Renewal Processes," J. Operations Res. Soc. Japan, Vol. 12, pp. 127-188.

[9] S. Osaki (1980), "A Two-Unit Parallel Redundant System with Bivariate Exponential Lifetimes," Microelectron. Reliab., Vol. 20, pp. 521-523.

[10] S. Osaki and T. Asakura (1970), "A Two-Unit Standby Redundant System with Repair and Preventive Maintenance," J. Applied Probability, Vol. 7, pp. 641-648.

[11] S. Osaki and T. Nakagawa (1976), "Bibliography for Reliability and Availability of Stochastic Systems," IEEE Trans. Reliability, Vol. R-25, pp. 284-287.

[12] W.P. Pierskalla and J.A. Voelker (1976), "A Survey of Maintenance Models: The Control and Surveillance of Deteriorating Systems," Naval Res. Logistics Q., Vol. 23, pp. 353-388.

[13] Y.S. Sherif and M.L. Smith (1981), "Optimal Maintenance Models for Systems Subject to Failure - A Review," Naval Res. Logistics Q., Vol. 28, pp. 47-74.

CHAPTER 6

STOCHASTIC MODELS FOR FAULT-TOLERANT COMPUTING SYSTEMS

6.1 Introduction

The remarkable progress of advanced computer technology enables us to make large, highly reliable computing systems for military and commmercial missions such as air traffic control, banking, seat reservations, communications, and databases. If such a system breaks down, there is economic damage and social confusion. It is, therefore, of great importance that such systems be very reliable.

The concept of fault-tolerance was introduced in the late 1960s. Avizienis (1976) defined a fault-tolerant system and a partially fault-tolerant (failsoft or gracefully degrading) system.

In general, highly reliable computing systems can be achieved by redundancy and/or maintenance techniques. The redundancy techniques that protect computing systems againist operational faults have three different forms: hardware (additional components), software (special programs), and time (repetition of operations) (see Avizienis (1976)). This chapter discusses stochastic models for fault-tolerant computing systems from the viewpoint of hardware redundancy.

Beaudry (1978) proposed the following four types of redundant computing systems with several processors:

(1) Massive Redundant System.
(2) Standby Redundant System.
(3) Hybrid Redundant System.
(4) Gracefully degrading System.

These redundant systems have their own characteristics of architecture, performance, and reliability as follows:

1. Massive redundant systems use techniques such as triple-modular redundancy, N-modular redundancy, or self-purging redundancy. They execute the same tasks on each equivalent unit (processor) and vote the output for improving the output information. A dual system, which is composed of two identical units, executes the same tasks on each unit and checks the output; it is one of the simplest massive redundant systems.

2. Standby redundant systems execute tasks on their active units. Upon detection of the failure of an active unit, the unit attempt to replace the faulty unit with a spare unit. A duplex system, which is composed of an active unit and a spare unit, is one of the simplest standby redundant systems.

3. Hybrid redundant systems are composed of a massive redundant (or gracefully degrading) core with spares to replace fault units.

4. Gracefully degrading systems use all units to execute tasks, i.e., all failure-free units are active. When a unit failure is detected, these systems attempt to reconfigure to a system with one fewer units. A multi-system (multiple, multiple-processor or multi-processor system) is one such gracefully degrading system.

Several reliability measures have been proposed and adopted to evaluate a system (see Section 4.1). In fault-tolerant computing, the following measures have been directly used:

(i) Reliability, R(t).
(ii) Mean Time to First Failure, MTFF.
(iii) Mean Time Between Failures, MTBF.
(iv) Mean Down Time, MDT.
(v) Pointwise and limiting Availabilities, A(t), A.

The measures (iii) - (v) apply only to a repairable system; however, measures (i) - (ii) can be applied to nonrepairable systems as well.

For computing systems, the trade-off between the reliability and performance is important. Under the same budget constraints of two different computing system configurations, the higher the reliability is, the lower the performance is. For instance, the performance of a dual system is less than that of a simplex system if we consider a simplex system and the corresponding dual system composed of two similar units (processors). Therefore, we are interested in gracefully degrading systems that balance reliability with performance. This chapter mainly considers gracefully degrading systems from the viewpoint of generalized reliability measures. The reliability measures described above (i) - (v) are not adequate for evaluating computing systems since performance is not taken into account. Beaudry (1978) introduced the following performance-related reliability measures:

(vi) Computation reliability, $R^*(t, T)$.

(vii) Mean Computation to First Failure, MCFF.
MCFF = the expected amount of computation available on a system before its first failure, given an initial system state.

(viii) Computation thresholds, t_T, T_t.

(ix) Computation availabilities, $A_c(t)$, A_c.

(x) Capacity threshold, t_c.
t_c = time at which the computation availability reaches a specific value.

Beaudry (1978) called (vii) the Mean Computation Before Failure instead of MCFF. We introduce the new measure:

(xi) Mean Computation Between Failures, MCBF.
MCBF = the expected amount of computation available on a system between two successive failures.

The performance-related reliability measures (vi) - (xi) do not take account of computational demands. Gay and Ketelsen (1979) proposed some performance measures with computational demands:

(xii) Expected steady-state throughput.
(xiii) Throughput availability.
(xiv) Lost throughput.

They discussed a two-processor system by making all the exponential assumptions, derived the measures above, and compared three possible methods of assigning incoming transactions to processors.

Nakamura and Osaki (1984) discussed a multi-processor system of two independent and identically distributed lifetime distributions of units under the generalized assumptions. They derived the exact and approximate formulae by applying Markov renewal processes, and showed that the approximate formulae have sufficient precision in practice. We will apply their results to our models developed below in the following section.

6.2 Multi-Processor Systems

We consider a multi-processor system with a buffer (see Fig. 6.2.1). The multi-processor system consists of two identical processors. The arriving jobs can be stored in the buffer and assigned to each processor which can process a single job at a time if it is available.

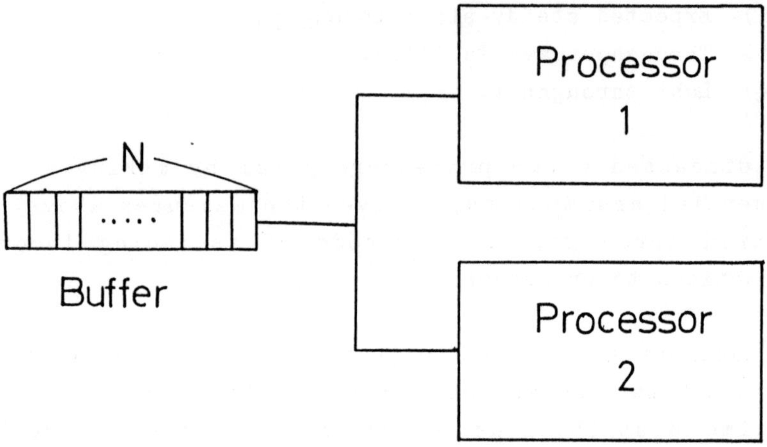

Fig. 6.2.1. A multi-processor system.

The lifetime of each unit is exponential with parameter λ_1, and the repair time distribution of the failed unit is an arbitrary distribution $G_1(t)$ with a finite mean $1/\mu_1$. The repair facility is a single, which means that there exists a queue and the repair discipline will be mentioned later. When one of two operating units is failed or the failed unit is repaired while another unit is operating, the automatic reconfiguration is made. Then we introduce the coverage, i.e., the probability of the correct automatic reconfiguration. If the automatic reconfiguration is made unsuccessfully, the manual reconfiguration is made, and its reconfiguration time is arbitrarily distributed. Let $G_2(t)$ denote the manual reconfiguration time distribution with a finite mean $1/\mu_2$.

Next, consider the failure and the repair of the buffer. Let N denote the storage capacity of the buffer, i.e., the buffer can store N Jobs. We assume that the buffer failure time is exponentially distributed with parameter λ_2 when $N = 1$. Then, if the storage capacity of the buffer is N, the buffer failure rate is $N\lambda_2$. The same assumptions have been made by Meyer (1982). Once the buffer is failed, the system can loose its functioning immediately. The repair discipline of the failed buffer is as follows: When the buffer is failed while

i) both units is operating, the failed buffer is repaired immediately;

ii) one of two units is under repair, the failed buffer joins a queue for repair. Then the repaired unit has to wait for its operation until the repair completion of the failed buffer.

The repair time distribution of the buffer is an arbitrary distribution $G_3(t)$ with a finite mean $1/\mu_3$.

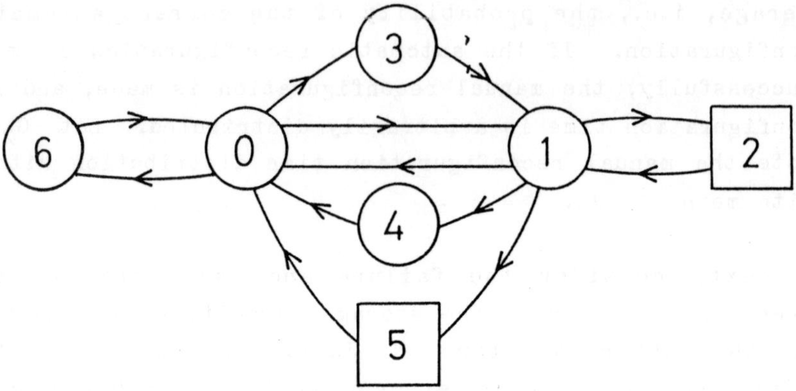

Fig. 6.2.2. A state transition diagram of the multiprocessor system.

Under the above assumptions, we define the following states of the system:

State 0: Both units and a buffer are operating.

State 1: One unit is failed while it is in state 0. The reconfiguration is completed and the repair of the failed unit starts.

State 2: One unit is failed while another unit is under repair (system failure).

State 3: One processor is failed while it is in state 0. The automatic reconfiguration is made unsuccessfully and the manual one starts (system failure).

State 4: The repair of the failed unit is completed while it is in state 1. The automatic reconfiguration is made unsuccessfully and the manual one starts (system failure).

State 5: A buffer is failed while it is in state 1 (system failure).

State 6: A buffer is failed while it is in state 0 (system failure).

We assume that available units (processors) and/or a buffer does not fail when it is in the system failure states 2, 3, ..., 6. Fig. 6.2.2 shows the state transition diagram, where the number circled denotes a regeneration point and the number squared a non-regeneration point.

Finally, consider the arriving jobs. We assume that the processing discipline of the job is 'first come, first served', and the arriving jobs form an M/M queueing system (see Section 2.5), where the arrival rate is λ_T and the processing rate is μ_T. Taking account of the number of the available units and the storage capacity of the buffer, we can consider that an M/M/2/N+2 queueing system is formed in state 0 and an M/M/1/N+1 queueing system is formed in state 1.

To analyze the model mentioned above, we apply the Markov renewal processes with non-regeneration points developed in Section 3.5 and queueing theory.

First, we derive the Laplace-Stieltjes transforms

$Q^*_{ij}(s)$ and $Q^{(k)*}_{ij}(s)$ of the one-step and two-step transition probabilities $Q_{ij}(t)$ and $Q^{(k)}_{ij}(t)$ by applying the Markov renewal process with non-regeneration points. Let c denote the coverage. We have

(6.2.1) $$Q^*_{01}(s) = c \cdot 2\lambda_1/(s + 2\lambda_1 + N\lambda_2) ,$$

(6.2.2) $$Q^*_{03}(s) = (1 - c)2\lambda_1/(s + 2\lambda_1 + N\lambda_2) ,$$

(6.2.3) $$Q^*_{06}(s) = N\lambda_2/(s + 2\lambda_1 + N\lambda_2) ,$$

(6.2.4) $$Q^*_{60}(s) = G^*_3(s) ,$$

(6.2.5) $$Q^*_{10}(s) = c \cdot G^*_1(s + \lambda_1 + N\lambda_2) ,$$

(6.2.6) $$Q^*_{14}(s) = (1 - c)G^*_1(s + \lambda_1 + N\lambda_2) ,$$

(6.2.7) $$Q^*_{31}(s) = Q^*_{40}(s) = G^*_2(s),$$

(6.2.8) $$Q^*_{12}(s) = \lambda_1[1 - G^*_1(s+\lambda_1+N\lambda_2)]/(s + \lambda_1 + N\lambda_2) ,$$

(6.2.9) $$Q^{(2)*}_{11}(s) = \lambda_1[G^*_1(s) - G^*_1(s+\lambda_1+N\lambda_2)]/(\lambda_1 + N\lambda_2) ,$$

(6.2.10) $$Q^*_{15}(s) = N\lambda_2[1 - G^*_1(s+\lambda_1+N\lambda_2)]/(s + \lambda_1 + N\lambda_2) ,$$

(6.2.11)
$$Q^{(5)*}_{10}(s) = N\lambda_2[G^*_1(s) - G^*_1(s+\lambda_1+N\lambda_2)]G^*_3(s)/(\lambda_1 + N\lambda_2),$$

where $G^*_i(s)$ $(i = 1, 2, 3)$ denote the Laplace-Stieltjes transforms of $G_i(t)$, respectively.

The unconditional means ξ_i $(i = 0, 1, 3, 4, 6)$ neglecting the non-regeneration points 2 and 5 (see Section 3.5) are given by

(6.2.12) $\xi_0 = 1/(2\lambda_1 + N\lambda_2)$,

(6.2.13) $\xi_1 = 1/\mu_1 + N\lambda_2[1 - G_1^*(\lambda_1+N\lambda_2)]/[\mu_3(\lambda_1 + N\lambda_2)]$,

(6.2.14) $\xi_3 = \xi_4 = 1/\mu_2$,

(6.2.15) $\xi_6 = 1/\mu_3$,

We also obtain the unconditional means η_i (i = 0, 1, ..., 6) not neglecting the non-regeneration points 2 and 5 (see Section 3.5):

(6.2.16) $\eta_i = \xi_i$ (i = 0, 3, 4, 6),

(6.2.17) $\eta_1 = [1 - G_1^*(\lambda_1+N\lambda_2)]/(\lambda_1 + N\lambda_2)$,

(6.2.18) $\eta_2 = \lambda_1\{1/\mu_1 - [1 - G_1^*(\lambda_1+N\lambda_2)]/(\lambda_1+N\lambda_2)\}/(\lambda_1+N\lambda_2)$,

(6.2.19)
$$\eta_5 = N\lambda_2\{1/\mu_1 + [1 - G_1^*(\lambda_1+N\lambda_2)]\cdot[1/\mu_3 - 1/(\lambda_1 + N\lambda_2)]\}/(\lambda_1 + N\lambda_2) .$$

In terms of $Q_{ij}^*(0)$, $Q_{ij}^{(k)*}(0)$ and ξ_i, we can obtain the mean recurrence times ℓ_{ii} for all the regeneration points (i = 0, 1, 3, 4, 6):

(6.2.20) $\ell_{00} = D/(1 - Q_{11}^{(2)*})$,

(6.2.21) $\ell_{11} = D/(1 - Q_{06}^*)$,

(6.2.22)
$$\ell_{33} = D/[1 - Q_{11}^{(2)*} - Q_{06}^* - Q_{11}^*(Q_{10}^* + Q_{14}^* + Q_{10}^{(5)*}) + Q_{11}^{(2)*}Q_{06}^*] ,$$

(6.2.23)
$$\ell_{44} = D/[1 - Q_{11}^{(2)*} - Q_{06}^{*} - (Q_{01}^{*} + Q_{03}^{*})(Q_{10}^{*} + Q_{10}^{(5)*}) \cdot + Q_{11}^{(2)*}Q_{06}^{*}] ,$$

(6.2.24) $$\ell_{66} = D/[1 - Q_{11}^{(2)*} - (Q_{01}^{*} + Q_{03}^{*})(Q_{10}^{*} + Q_{14}^{*} + Q_{10}^{(5)*}],$$

where $Q_{ij} \equiv Q_{ij}(0)$, $Q_{ij}^{(k)*} \equiv Q_{ij}^{(k)*}(0)$ and

(6.2.25) $$D = (1 - Q_{11}^{(2)*})(\xi_0 + Q_{03}^{*}\xi_3 + Q_{06}^{*}\xi_6) + (Q_{03}^{*} + Q_{01}^{*})\xi_1 + (Q_{03}^{*} + Q_{01}^{*})Q_{14}^{*}\xi_4 .$$

From the above results, we can derive the limiting probabilities for all the states and the expected number of visits to state i $(i = 0, 1, 2, 3, 4, 5, 6)$ per unit time in the steady- state. Let ${}_iP$ denote the limiting probability for state i. Then we have

(6.2.26) $${}_iP = \eta_i/\ell_{ii} \qquad (i = 0, 1, 3, 4, 6),$$

(6.2.27) $${}_iP = \eta_i/\ell_{11} \qquad (i = 2, 5).$$

The expected numbers of visits to state i per unit time in the steady-state, M_i, are given by

(6.2.28) $$M_i = 1/\ell_{ii} \qquad (i = 0, 1, 3, 4, 6),$$

(6.2.29) $$M_i = Q_{1i}^{*}/\ell_{ii} \qquad (i = 2, 5).$$

Next we consider the queue of the jobs in states 0 and 1. Recall that an M/M/2/N+2 queueing system is formed in state 0. Let P_j denote the steady-state probability that the number of the jobs in the system, i.e., the stored jobs in the buffer plus the processing jobs, is j. Then P_j (j = 0, 1, ..., N+2) are given by

(6.2.30) $$P_0 = [1 + \rho + \rho^2/2 + 2\sum_{k=3}^{N+2} \rho^k]^{-1} ,$$

(6.2.31) $$P_1 = \rho P_0 ,$$

(6.2.32) $$P_j = 2\rho^j P_0 \quad (j = 2, 3, ..., N+2) ,$$

where $\rho = \lambda_T/\mu_T$ (see Kleinrock (1975)). Also, we recall that the arriving jobs forms an M/M/1/N+1 queueing system in state 1. Let P_m^* denote that the steady-state probability that the number of the jobs in the system is m. We have

(6.2.33) $$P_m^* = (1 - \rho)\rho^m/(1 - \rho^{N+2}) \quad (\rho \neq 1) ,$$

(6.2.34) $$P_m^* = 1/(N + 2) \quad (\rho = 1) ,$$

where m = 0, 1, 2, ..., N+1 (see Kleinrock(1975)).

Let ${}_iP_j$ denote the steady-state probability that the number of the jobs in the system is j in state i. From the assumption that the behaviors of the failure-repair and of the demand of the jobs are independent, ${}_iP_j$ are derived by using ${}_iP$, P_j and P_j:

(6.2.35) $${}_0P_i = {}_0P \cdot P_j \quad (j = 0, 1, 2, ..., N+2),$$

(6.2.36) $${}_1P_j = {}_1P \cdot P_j^* \quad (j = 0, 1, 2, ..., N+2).$$

6.3 Performance/Reliability Measures and Numerical Examples

We adopt the following measures to evaluate the multiprocessor system from the viewpoints of reliability and performance:

(1) The steady-state availability A_v, which is the probability that the system is operating in the steady-state;

(6.3.1) $$A_v = {}_0P + {}_1P.$$

(2) The MTBF, which is the mean time between failures;

(6.3.2) $$MTBF = A_v / (\sum_{i=2}^{6} M_i) .$$

(3) The computation availability A_c, which is the expected value of computation capacity of the system in the steady-state (see Beaudry (1978));

(6.3.3) $$A_c = 2\mu_T \cdot {}_0P + \mu_T \cdot {}_1P.$$

(4) The expected system throughput T_p, which was introduced by Gay and Ketelsen (1979);

(6.3.4) $$T_p = \sum_i \sum_j \mu_{ij} \cdot {}_iP_j ,$$

where μ_{ij} is the throughput for state i with j jobs. In this model,

(6.3.5) $\mu_{ij} = 2\mu_T \quad (i = 0;\ j = 2, 3, \ldots, N+2),$

(6.3.6) $\mu_{ij} = \mu_T \quad (i = 0;\ j = 1:\ i = 1;\ j = 1, 2, \ldots, N+2).$

Then we have

(6.3.7) $$T_p = 2\mu_T \cdot \sum_{j=2}^{N+2} {}_0P_j + \mu_T({}_0P_1 + \sum_{j=1}^{N+1} {}_1P_j) .$$

(5) The expected number of lost jobs by the failure per unit time L_J, which is the expected number of jobs lost by failures of processors, buffer, or unsuccessfully automatic reconfiguration per unit time;

(6.3.8) $$L_J = \sum_i \sum_j m_{ij}\, n_{ij}$$

where m_{ij} and n_{ij} are the expected numbers of transitions from state i to state j in the steady-state and of the lost jobs by its transition, respectively. Gay and Ketelsen (1979) introduced the similar measures, which is, however, applied only to the Markov model. For example, we assume that if the system failure takes place, all the jobs in the system are lost, and that if only one processor is failed and another processor is still operating, a processing job in the failed processor is lost. Then we have

(6.3.9) $$L_J = M_0 Q_{01}^{*}(1 - P_0 - P_1/2) + (\sum_{k=1}^{N+2} {}_kP_k)(M_3 + M_6) + (\sum_{k=1}^{N+1} kP_k^{*})(M_2 + M_4 + M_5) .$$

(6) The <u>expected number of lost jobs by the cancellation per unit time</u> C_J;

(6.3.10) C_J = (arrival rate of the jobs) × P{the arriving jobs can not be accepted to the system}.

We assume that the arriving jobs cannot be accepted when the system is failed or the storage capacity of the buffer is full. Then we have

(6.3.11) $$C_J = \lambda_T(\sum_{i=2}^{6} {}_iP + {}_0P_{N+2} + {}_1P_{N+1}) .$$

The readers should notice the following identity between the expected system throughput T_p and C_J:

(6.3.12) $$T_p + C_J = \lambda_T ,$$

since the expected system throughput T_p represents the net arrival rate and C_J the cancelled arrival rate.

(7) The <u>total loss</u> W, which can be calculated by using L_J and C_J;

(6.3.13) $$W = \omega_1 L_J + \omega_2 C_J ,$$

where ω_1 and ω_2 are weights which characterize the lost jobs caused by the failure and by the cancellation, respectively.

We show the numerical examples of the above analytical results, where the repair time distributions $G_1(t)$, $G_3(t)$ and the manual reconfiguration time distribution $G_2(t)$ are the Erlang (gamma) distributions with shape parameter 2 (see Table 1.2.2):

$$(6.3.14) \qquad G_i(t) = 1 - (1 + 2\mu_i t)e^{-2\mu_i t} .$$

The plots depicted in the figures are obtained by assuming that $\lambda_1 = 10^{-3}$, $\mu_1 = 0.05$, $\mu_2 = 0.05$ and $c = 0.999$, where the parameter λ_2 varies.

Fig. 6.3.1 shows the values of the steady-state availability A_V depending on the storage capacity of the buffer. We see from Fig. 6.3.1 that when N increases, A_V decreases except the case of $\lambda_2 = 0$. That is, the increase of N causes the increase of the number of the buffer failures. The steady-state availability has no any impact on system performance since it is independent of the parameters λ_T and μ_T. That is, the steady-state availability cannot contribute to any system performance even if the number of buffer capacity increases. The tendency of the MTBF is quite similar to that of the steady-state availability.

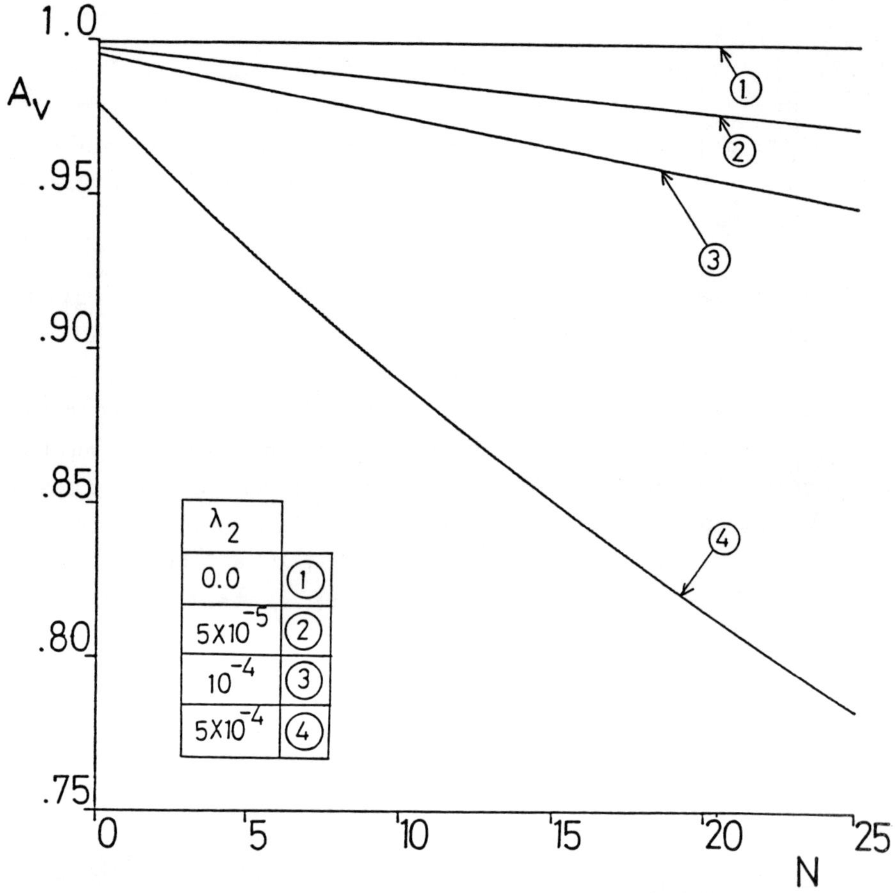

Fig. 6.3.1. The curves of the steady-state availability A_v depending on the storage capacity of the buffer N.

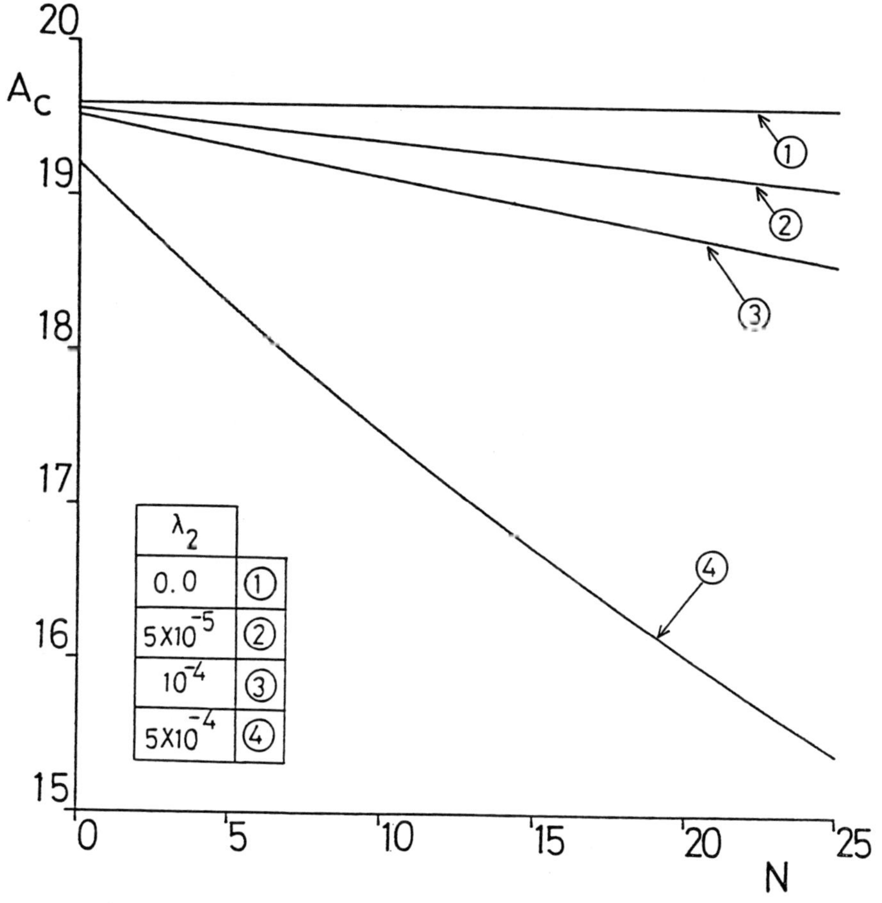

Fig. 6.3.2. The curves of the computation availability A_c depending on the storage capacity of the buffer N.

Fig. 6.3.2 shows the values of the computation availability A_c depending on the storage capacity of the buffer, where $\mu_T = 10$. We can understand from this figure that the computation availability decreases as N increases as well as the steady-state availability. Therefore, the computation availability has not any impact on whether there exist the computational demands or not, since it is independent of the parameter λ_T, the arrival rate of the jobs.

Fig. 6.3.3 shows the values of the expected system throughput T_p depending on the storage capacity of the buffer, where $\mu_T = 10$ and $\lambda_T = 16$. From this figure we obtain the following: When $\lambda_2 = 0$, i.e., the buffer is assumed not to be failed, the expected system throughput increases as N increases. However, the optimum N can be obtained by maximizing the expected system throughput, where $\lambda_2 > 0$. For example, the optimum N is 7 if $\lambda_2 = 5\times10^{-4}$. Since this measure represents the trade-off between the increase of performance and the decrease of reliability, we should maximize it.

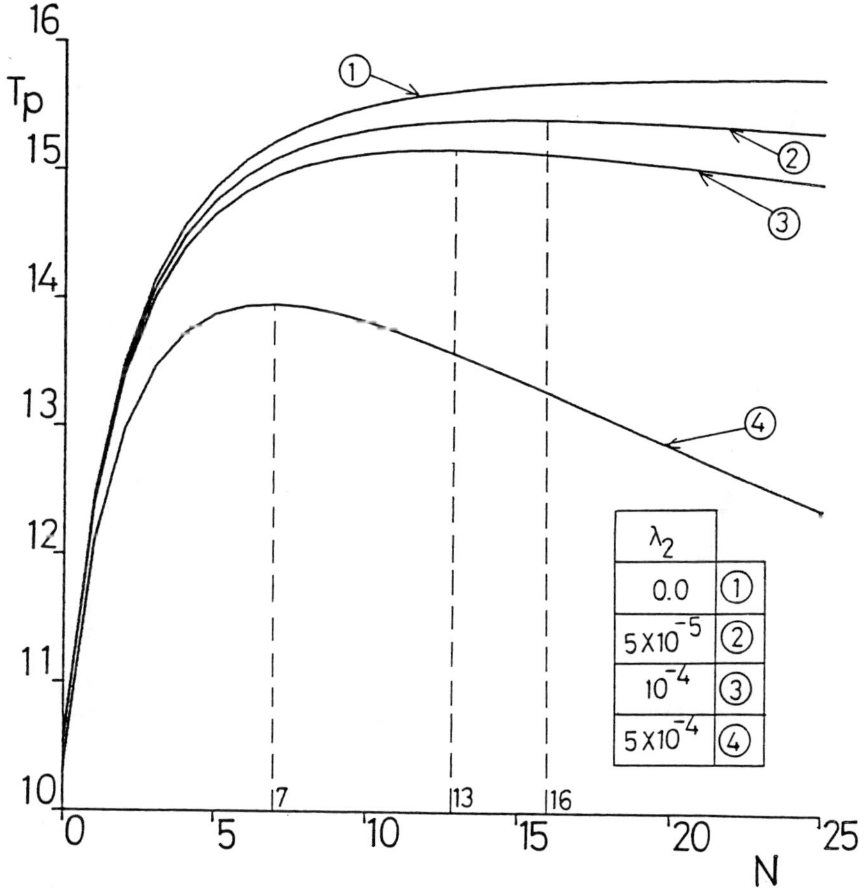

Fig. 6.3.3. The curves of the expected system throughput T_p depending on the storage capacity of the buffer N.

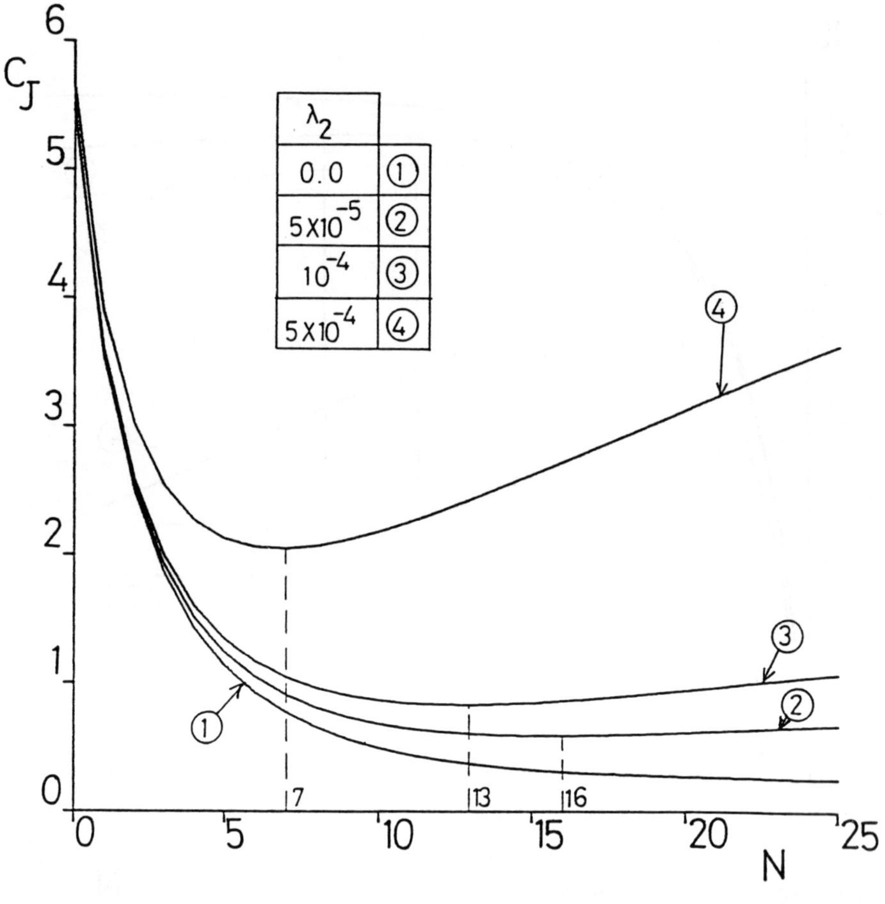

Fig. 6.3.4. The curves of the expected number of lost jobs by cancellation per unit time, C_J, depending on the storage capacity of the buffer N.

Fig. 6.3.4 shows the values of the expected number of lost jobs by the cancellation per unit time, C_J, depending on the storage capacity of the buffer, where $\mu_T = 10$ and $\lambda_T = 16$. Comparing Fig. 6.3.3 with Fig. 6.3.4, we see that the tendency of C_J is contrary to that of the expected system throughput. That is, C_J represents the trade-offs between performance and reliability, and gives the optimum N to be minimized. If λ_2 are equal to 5×10^{-4}, 10^{-4}, 5×10^{-5}, the optimum N which minimize C_J are 7, 13, and 16, respectively, and are equal to ones which maximize the expected system throughput. This fact can be easily understood from the identity (6.3.12). That is, maximizing T_p is equivalent to minimizing C_J.

Table 6.3.1. The values of the expected number of lost jobs by failure per unit time, L_J.

	$\lambda_2 = 10^{-5}$		$\lambda_2 = 10^{-4}$		$\lambda_2 = 5\times10^{-4}$	
	$\rho = 1.0$	$\rho = 1.6$	$\rho = 1.0$	$\rho = 1.6$	$\rho = 1.0$	$\rho = 1.6$
	$(\times10^{-2})$	$(\times10^{-2})$	$(\times10^{-2})$	$(\times10^{-2})$	$(\times10^{-2})$	$(\times10^{-2})$
L = 0	.080	.108	.094	.126	.155	.207
L = 5	.117	.183	.200	.353	.543	1.060
L = 10	.136	.235	.290	.635	.894	2.212
L = 15	.155	.287	.384	.937	1.237	3.385
L = 20	.175	.338	.484	1.242	1.578	4.482
L = 25	.196	.389	.591	1.547	1.916	5.499

Finally, Table 6.3.1 shows the expected number of lost jobs by the failure per unit time L_J, where the values of the parameters λ_1, μ_1, μ_2, μ_3 and c are same as ones used in the preceding figures. We see from this table that L_J increases as the storage capacity of the buffer, the traffic density ρ $(= \lambda_T/\mu_T)$ and the value of the parameter λ_2 increase.

In this chapter we have discussed several performance/reliability measures to evaluate a multiprocessor system consisting of two processors and a buffer. We have assumed that the processor and the buffer are repairable, and considered the arrival, processing and queues of the jobs in order to take account of the system performance. Applying the Markov renewal processes with non-regeneration points and the queueing theory, we have derived the performance/reliability measures such as the steady-state availability, the MTBF, the computation availability, the expected system throughput and the expected numbers of lost jobs. Furthermore, we have discussed which measure represents the trade-offs between performance and reliability explicitly.

Numerical examples have shown the influence of the storage capacity of the buffer on the performance/reliability measures. We can conclude the following: The reliability measures such as the steady-state availability and the MTBF represent only the decrease of reliability by the increase of the buffer capacity. The computation availability, which is called the performance-related reliability measure, has the same tendency, since it takes account of the arrival rate of the jobs. But the expected system throughput and the expected number of lost jobs by the cancellation represent the performance and reliability simultaneously, and give the optimum storage capacity of the buffer balancing the trade-offs between performance and reliability.

Bibliography and Comments

Throughout all the sections: Siewiorek and Swarz (1982) edited the present status of fault-tolerant computing systems. Avizienis (1976) discussed fault-tolerant systems in general. Osaki and Nishio (1980) discussed stochastic models for fault-tolerant computing systems by applying Markov renewal processes with non-regeneration points.

Section 6.1: We are following the papers by Beaudry (1978), Gay and Ketelsen (1979), and Osaki (1984). Meyer (1982) proposed 'performability'.

Sections 6.2 and 6.3: We are following the papers by Osaki and Nakamura (1984), and Nakamura and Osaki (1984). For the results of M/M queueing systems, see Kleinrock (1975).

[1] A. Avizienis (1976), "Fault-Tolerant Systems," IEEE Trans. Computers, Vol. C-25, pp. 1304-1312.

[2] M.D. Beaudry (1978), "Performance-Related Reliability Measures for Computing Systems," IEEE Trans. Computers,

Vol. C-27, pp. 540-547.

[3] F.A. Gay and M.L. Ketelsen (1979), "Performance Evaluation for Gracefully Degrading Systems," Proc. Int. Symp. Fault-Tolerant Computing, pp. 51-58.

[4] L. Kleinrock (1975), Queueing Systems, Volume I: Theory, John Wiley and Sons, New York.

[5] J.F. Meyer (1982), "Closed-Form Solutions of Performability," IEEE Trans. Computers, Vol. C-31, pp. 648-657.

[6] M. Nakamura and S. Osaki (1984), "Performance/Reliability Evaluation for Multi-Processor Systems with Computational Demands," International J. Systems Sci., Vol. 15, pp. 95-105.

[7] S. Osaki (1984), "Performance/Reliability Measures for Fault-Tolerant Computing Systems," IEEE Trans. Reliability, Vol. R-33, pp. 268-271.

[8] S. Osaki and M. Nakamura (1984), "Performance/Reliability Modeling for Multi-Processor Systems with Computational Demands," S. Osaki and Y. Hatoyama (eds.), Stochastic Models in Reliability Theory, Springer-Verlag, Berlin.

[9] S. Osaki and T. Nishio (1980), Reliability Evaluation of Some Fault-Tolerant Computer Architectures, Springer-Verlag, Berlin.

[10] D.P. Siewiorek and R.S. Swarz (1982), The Theory and Practice of Reliable System Design, Digital Press, Bedford, Massachusetts.

APPENDIX A

LAPLACE-STIELTJES TRANSFORMS

Let F(t) be a well-defined function of t specified for $t \geq 0$ and s be a complex number. If the following Stieltjes integral

(A.1) $$F^*(s) = \int_0^\infty e^{-st} dF(t)$$

converges on some s_0, the Stieltjes integral (A.1) converges on any s such that $Re(s) > Re(s_0)$. The integral (A.1) is called the Laplace-Stieltjes transform of F(t). If the real function F(t) can be expressed in terms of the following integral:

(A.2) $$F(t) = \int_0^t dF(x) = \int_0^t f(x)dx,$$

then

(A.3) $$F^*(s) = \int_0^\infty e^{-st} dF(t) = \int_0^\infty e^{-st} f(t)dt,$$

which is the Laplace transform of f(t).

Noting that F(t) is one-to-one correspondent with $F^*(s)$ (see Theorem (1.2.28) and its following discussion),

$F(t)$ can be uniquely specified by $F^*(s)$. The inversion formula for obtaining $F(t)$ from $F^*(s)$ can be given by

(A.4) $$F(t) = \lim_{c\to\infty} \frac{1}{2\pi i}\int_{b-ic}^{b+ic} \frac{e^{st}}{s} F^*(s)ds ,$$

where $i = \sqrt{-1}$ is an imaginary unit, $b > \max(\sigma, 0)$ and σ is a radius of convergence.

The following two theorems are well-known and of great use as the limit theorems for the Laplace-Stieltjes transform $F^*(s)$ of $F(t)$.

(A.5) Theorem (An Abelian Theorem) If for some non-negative number α,

(A.6) $$\lim_{t\to\infty} \frac{F(t)}{t^\alpha} = \frac{C}{\Gamma(\alpha+1)} ,$$

then

(A.7) $$\lim_{s\to+0} s^\alpha F^*(s) = C,$$

where $\Gamma(n) = \int_0^\infty e^{-x}x^{n-1}dx$ is a gamma function of order n.

(A.8) Theorem (A Tauberian Theorem) IF $F(t)$ is non-decreasing and the Laplace-Stieltjes transform

(A.9) $$F^*(s) = \int_0^\infty e^{-st}dF(t)$$

converges for $Re(s) > 0$, and if for some non-negative number α,

(A.10) $$\lim_{s\to+0} s^{\alpha} F^{*}(s) = C,$$

then

(A.11) $$\lim_{t\to\infty} F(t)/t^{\alpha} = C/\,\Gamma(\alpha + 1).$$

Table A.1 shows the general properties of the Laplace-Stieltjes transforms. Table A.2 also shows the Laplace-Stieltjes transform formulae of the typical functions. Applying Tables A.1 and A.2, we can obtain $F(t)$ from the Laplace-Stieltjes transform $F^{*}(s)$ and vice versa.

(A.12) Example The Laplace-Stieltjes transform of the exponential distribution having parameter λ (see Table 1.2.2) is given by

(A.13) $$F^{*}(s) = \int_0^\infty e^{-st} dF(t) = \int_0^\infty e^{-st} \lambda e^{-\lambda t} dt = \lambda/(s + \lambda).$$

(A.14) Example The Laplace-Stieltjes transform of the Poisson distribution (see Table 1.2.1) is given by

(A.15) $$F^{*}(s) = \int_0^\infty e^{-st} dF(t) = \sum_{x=0}^{\infty} e^{-sx} e^{-\lambda} \frac{\lambda^{x}}{x!} = \exp[-\lambda(1-e^{-s})].$$

(A.16) Example As shown in Example (1.3.40), if $X_1, X_2, \ldots, X_n$ are independent and identically distributed exponential random variables with parameter λ, then the Laplace-Stieltjes transform of the distribution $P\{S_n \le t\}$, where $S_n = X_1 + X_2 + \ldots + X_n$, is given by

(A.17) $F^{*}(s) = [\lambda/(s+\lambda)]^{n}.$

Applying the formulae in Tables A.1 and A.2, we have

(A.18) $$F(t) = \int_0^t e^{-\lambda x} \frac{\lambda(\lambda x)^{n-1}}{(n-1)!} dx$$

which is a gamma distribution of order n (see Table 1.2.2).

Reference

[1] D.W. Widder (1946), The Laplace Transform, Princeton University Press, Princeton, New Jersey.

Table A.1. Table of general properties of the Laplace-Stieltjes transforms.

$F(t)$	$F^*(s) = \int_0^\infty e^{-st} dF(t)$
$F_1(t) + F_2(t)$	$F_1^*(s) + F_2^*(s)$
$aF(t)$	$aF^*(s)$
$F(t-a) \quad (a > 0)$	$e^{-sa}F^*(s)$
$F(at) \quad (a > 0)$	$F^*(s/a)$
$e^{-at}F(t) \quad (a > 0)$	$[s/(s + a)]F^*(s+a)$
$F'(t)$	$sF^*(s) - F(0)$
$tF'(t)$	$-sF^*(s)$
$[t(d/dt)]^n F(t)$	$[-s(d/ds)]^n F^*(s)$
$\int_0^t F(x)dx$	$(1/s)F^*(s)$
$\int_0^t \ldots \int_0^t F(t)(dt)^n$	$(1/s^n)F^*(s)$
$\lim_{t\to+0} F(t)$	$\lim_{s\to\infty} F^*(s)$
$\lim_{t\to\infty} F(t)$	$\lim_{s\to+0} F^*(s)$

Table A.2. Table of special Laplace-Stieltjes transforms.

$F(t)$	$F^*(s) = \int_0^\infty e^{-st}dF(t)$
$\delta(t-a)$ $(a > 0)$	se^{-sa}
$1(t-a)$ $(a > 0)$	e^{-sa}
$1(t)$	1
t	$1/s$
t^n	$n!/s^n$
t^α $(\alpha > -1)$	$\Gamma(\alpha+1)/s^\alpha$
$e^{-\alpha t}$ $(\alpha > 0)$	$s/(s + \alpha)$
$te^{-\alpha t}$ $(\alpha > 0)$	$s/(s + \alpha)^2$
$t^n e^{-\alpha t}$ $(\alpha > 0)$	$n!s/(s + \alpha)^{n+1}$
$t^\beta e^{-\alpha t}$ $(\alpha > 0, \beta > -1)$	$s\Gamma(\beta+1)/(s + \alpha)^{\beta+1}$
$\cos \alpha t$	$s^2/(s^2 + \alpha^2)$ $(Re(s) > \|\alpha\|)$
$\sin \alpha t$	$s\alpha/(s^2 + \alpha^2)$ $(Re(s) > \|\alpha\|)$
$\cosh \alpha t$	$s^2/(s^2 - \alpha^2)$ $(Re(s) > \|\alpha\|)$
$\sinh \alpha t$	$s\alpha/(s^2 - \alpha^2)$ $(Re(s) > \|\alpha\|)$
$\log t$	$-\gamma - \log s$ †

† $\gamma = 0.57721\cdots$, Euler's constant.

APPENDIX B

SIGNAL-FLOW GRAPHS

A signal-flow graph is a graphical diagram composed of directed branches and nodes. As shown in Section 3.4, a state transition diagram among states in a Markov renewal process can be regarded as a signal-flow graph, where each node corresponds to each state and each branch gain corresponds to each Laplace-Stieltjes transform of each one-step transition probability or mass function if it exists.

In a signal-flow graph in general, some node is referred to as either a source or sink. A source is a node at which all branches are directed outward, and a sink is a node at which all branches are directed inward. However, as shown in Section 3.4, we might omit a source or sink if it is not confused.

Consider a signal-flow graph shown in Fig. B.1. The signal-flow graph is a state transition diagram for a 4-out-of-n: F system discussed by Osaki (1970), where each branch gain are given. We simply describe several terms for signal-flow graphs by illustrating Fig. B.1 as an example. Note in Fig. B.1 that state 0 is a source and state 3 is a sink. Each branch has each branch gain $Q^*_{ij}(s)$ if there exists a one-step transition probability $Q^*_{ij}(s)$ from state i to state j (i, j = 0, 1, 2, 3).

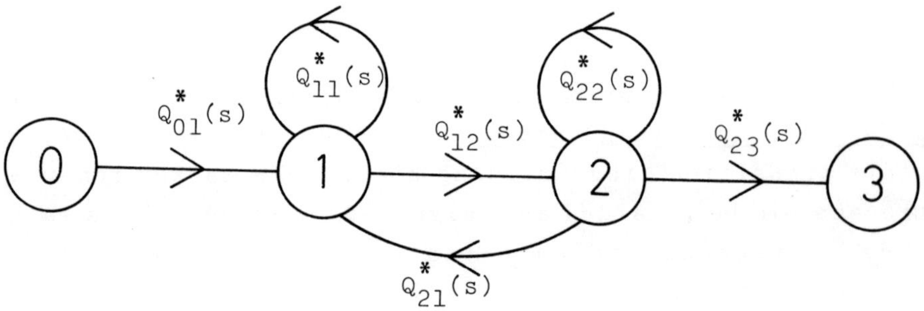

Fig. B.1. A signal-flow graph for obtaining the overall transmittance from source 0 to sink 3.

In Fig. B.1, there are two possible paths from state 0 to state 3, where the two possible paths are $0 \rightarrow 1 \rightarrow 2 \rightarrow 3$ and $0 \rightarrow 1 \rightarrow 3$. A feedback loop (or simply loop) is a path which starts that no node is traversed more than once. In Fig. B.1, there are 4 loops, i.e., $0 \rightarrow 1 \rightarrow 0$, $1 \rightarrow 1$, $1 \rightarrow 2 \rightarrow 1$, and $2 \rightarrow 2$. In particular, if a loop which contains only a single node is called a self-loop. In Fig. B.1, the loops $1 \rightarrow 1$ and $2 \rightarrow 2$ are self-loops.

A feedback path is a path which contains at least one feedback loop. On the other hand, an open path or a forward path is a path which contains no feedback loops. The path gain or path transmittance is the product of all transmittances of the branches of the path. The path gain $0 \rightarrow 1 \rightarrow 2 \rightarrow 3$ is $Q^*_{01}(s)Q^*_{12}(s)Q^*_{13}(s)$ and the path gain $0 \rightarrow 1 \rightarrow 3$ is $Q^*_{01}(s)Q^*_{13}(s)$. Similarly, the open-path gain or the forward-path gain is the product of transmittances associated with an open path. The loop gain is the product of all transmittances associated with a loop. In Fig. B.1, the loop gains $0 \rightarrow 1 \rightarrow 0$, $1 \rightarrow 1$, $1 \rightarrow 2 \rightarrow 1$, and $2 \rightarrow 2$ are $Q^*_{01}(s)Q^*_{10}(s)$, $Q^*_{11}(s)$, $Q^*_{12}(s)Q^*_{21}(s)$, and $Q^*_{22}(s)$, respectively.

Two method can be considered to obtain the overall transmittance from a source to a sink. One method is to reduce a signal-flow graph composed of only the source and the sink. Another method is to apply Mason's gain formula, which can give the overall transmittance from the source to the sink. We just cite the latter method which is simple and straightforward to obtain the probabilistic quantities of interest in a Markov renewal process.

Two gains or transmittances are involved in the formulation of Mason's gain formula, i.e., (1) loop-gain (transmittance) and (2) forward- (open-) path gain (transmittance). These gains were defined previously. With these definitions, Mason's gain formula is

$$H = \frac{1}{\Delta} \sum_{i=1}^{n} T_i \Delta_i \ , \tag{B.1}$$

where H = the overall transmittance from the source to the sink,

T_i = the gain of the ith forward (open) path from the source to the sink,

n = the total number of forward paths from the source to the sink,

and

Δ = the signal-flow graph determinant,

which is defined as follows:

$$\Delta = 1 - \sum_{i} T_{\ell i} + \sum_{i,j} T_{\ell i} T_{\ell j} - \cdots \ , \tag{B.2}$$

where the $T_{\ell i}$ is the loop gain, and in each of the product summations in (B.2) only products of nontouching loops are included. The term <u>nontouching</u> implies which have no node in common, i.e., separated loops. Note that the sign is minus for a sum of products of an odd number of loop gains and plus otherwise.

Similarly, the symbol Δ_i is the determinant Δ after all loops which touch the T_i path at any node have been eliminated. For instance, if we consider the T_1 path: $0 \to 1 \to 2 \to 3$, all the loops touch the T_1 path. However, if we consider the T_2 path: $0 \to 1 \to 3$, there exists one loop $2 \to 2$ which never touches the T_2 path at any node. That is, the determinants are $\Delta_1 = 1$ and

$$\Delta_2 = 1 - Q^*_{22}(s), \tag{B.3}$$

which can be obtained from equation (B.2).

In Fig. B.1, the overall transmittance from state 0 to state 3 is given by

$$G^*_{03}(s) = \{Q^*_{01}(s)Q^*_{12}(s)Q^*_{23}(s) + Q^*_{01}(s)Q^*_{13}(s)[1 - Q^*_{22}(s)]\}/\Delta, \tag{B.4}$$

where

$$\Delta = 1 - Q^*_{01}(s)Q^*_{10}(s) - Q^*_{11}(s) - Q^*_{12}(s)Q^*_{21}(s) - Q^*_{22}(s) + Q^*_{01}(s)Q^*_{10}(s)Q^*_{22}(s) + Q^*_{11}(s)Q^*_{22}(s). \tag{B.5}$$

Many examples for applying Mason's gain formula have been shown in Section 3.4. It is advisable to the readers how to apply Mason's gain formula for many examples and compare the results by applying the conventional techniques for solving the simultaneous linear equations developed in Section 3.3.

References

An example shown in Fig. B.1 was cited by Osaki (1970). The development of Mason's gain formula was given by Mason (1956), Ruston and Bordogna (1966), and Chen (1976).

[1] W.K. Chen (1976), Applied Graph Theory - Graphs and Electrical Networks, North-Holland, Amsterdam.

[2] S.J. Mason (1956), "Feedback Theory - Further Properties of Signal-Flow Graphs," Proc. IRE, Vol. 44, pp. 920-926.

[3] S. Osaki (1970), "System Reliability Analysis by Markov Renewal Processes," J. Operations Res. Soc. Japan, Vol. 12, pp. 127-188.

[4] H. Ruston and J. Bordogna (1966), Electric Networks: Functions, Filters, Analysis, McGraw-Hill, New York.

INDEX

E

F

G

H

I

J

K

L

M

N

T

U

V

W